P9-EKS-676

Product Design

Edited by **Akiko Busch** and the **Editors of *Industrial Design* Magazine**

MEDICALEQUIPMENTMEDICALEQUIP
EQUIPMENTMEDICALEQUIPMENTMED

INDUSTRIALEQUIPMENTTRANSPORTATIONINDUSTRIALEQUIPMENTTRAN
INDUSTRIALEQUIPMENTTRANSPORTATIONINDUSTRIALEQUIPMENTTRAN

RECREATIONALSPORTSEQUIPMENTRECREATIONALSPORTSEQUIPMENTRECREATIONALSPORTSEQUIPMENTREC
EQUIPMENTRECREATIONALSPORTSEQUIPMENTRECREATIONALSPORTSEQUIPMENTRECREATIONALSPORTSEQU

TEXTILESTEXTILESTEXTILESTEXTILESTEXTILESTEXTILESTEXTILESTEXTILESTEXTILESTEXTILESTEXTILESTEXTILESTEXTILESTEXTILE
TEXTILESTEXTILESTEXTILESTEXTILESTEXTILESTEXTILESTEXTILESTEXTILESTEXTILESTEXTILESTEXTILESTEXTILESTEXTILESTEXTILE

DESIGNSFORTHEHANDICAPPEDDESIGNSFORTHEHANDICAPPEDDESIGNSFORTHEHANDICAPPEDDESIGNSFORTHEHANDICAPPEDDESIGNSFORTHEHANDICAPPEDDESIGNSFORTHEHANDIC
FORTHEHANDICAPPEDDESIGNSFORTHEHANDICAPPEDDESIGNSFORTHEHANDICAPPEDDESIGNSFORTHEHANDICAPPEDDESIGNSFORTHEHANDICAPPEDDESIGNSFORTHEHANDICAPPEDD

International
Award-Winning
Designs for the
Home and Office

Product Design

APPLIANCESHOUSEWARESTOOLSAPPLIANC
APPLIANCESHOUSEWARESTOOLSAPPLIANC

HOMEELECTRONICSENTERTAINMENTHOMEELECTRONICSENTERTAINMENTHOMEE
ENTERTAINMENTHOMEELECTRONICSENTERTAINMENTHOMEELECTRONICSENTER

LIGHTINGLIGHTINGLIGHTINGLIGHTINGLIGHTINGLIGHTINGLIGHTINGLIGHTINGLIGHTINGLIGHTINGLIGHTING
LIGHTINGLIGHTINGLIGHTINGLIGHTINGLIGHTINGLIGHTINGLIGHTINGLIGHTINGLIGHTINGLIGHTINGLIGHTING

CONTRACTRESIDENTIALFURNISHINGSCONTRACTRESIDENTIALFURNISHINGSCONTRACTRESIDENTIALFURNISHINGSCONTRACTCONTRACTRESIDENTI
RESIDENTIALFURNISHINGSCONTRACTRESIDENTIALFURNISHINGSCONTRACTRESIDENTIALFURNISHINGSCONTRACTRESIDENTRESIDENTIALFURNIS

INESSEQUIPMENTBUSINESSEQUIPMENTBUSINESSEQUIPMENTBUSINESSEQUIPMENTBUSINESSEQUIPMENTBUSINESSEQUIPMENTBUSINESSEQUIPMENTBUSINESSEQUIPMENTBUSINES
INESSEQUIPMENTBUSINESSEQUIPMENTBUSINESSEQUIPMENTBUSINESSEQUIPMENTBUSINESSEQUIPMENTBUSINESSEQUIPMENTBUSINESSEQUIPMENTBUSINESSEQUIPMENTBUSINES

DICALEQUIPMENTMEDICALEQUIPMENTMEDICALEQUIPMENTMEDICALEQUIPMENTMEDICALMEDICALEQUIPMENTMEDICALEQUIPMENTMEDICALEQUIPMENTMEDICALEQUIPMENTMEDICAL
UIPMENTMEDICALEQUIPMENTMEDICALEQUIPMENTMEDICALEQUIPMENTMEDICALEQUIPME EQUIPMENTMEDICALEQUIPMENTMEDICALEQUIPMENTMEDICALEQUIPMENTMEDICALEQUIPME

TIONINDUSTRIALEQUIPMENTTRANSPORTATIONINDUSTRIALEQUIPMENTTRANSPORTATIONINDUSTRIALEQUIPMENTTRANSPORTATIONINDUSTRIALEQUIPMENTTRANSPORTATIONINDU
TIONINDUSTRIALEQUIPMENTTRANSPORTATIONINDUSTRIALEQUIPMENTTRANSPORTATIONINDUSTRIALEQUIPMENTTRANSPORTATIONINDUSTRIALEQUIPMENTTRANSPORTATIONINDU

ONALSPORTSRECREATIONALSPORTSEQUIPMENTRECREATIONALSPORTSEQUIPMENTRECREATIONALSPORTSEQUIPMENTRECREATIONALSPORTSRECREATIONALSPORTSEQUIPMENTRECRE
RECREATIOEQUIPMENTRECREATIONALSPORTSEQUIPMENTRECREATIONALSPORTSEQUIPMENTRECREATIONALSPORTSEQUIPMENTRECREATIOEQUIPMENTRECREATIONALSPORTSEQUIP

LESTEXTILESTEXTILESTEXTILESTEXTILESTEXTILESTEXTILESTEXTILESTEXTILESTEXTILESTEXTILESTEXTILESTEXTILESTEXTILESTEXTILESTEXTILESTEXTILESTEXTILEST
LESTEXTILESTEXTILESTEXTILESTEXTILESTEXTILESTEXTILESTEXTILESTEXTILESTEXTILESTEXTILESTEXTILESTEXTILESTEXTILESTEXTILESTEXTILESTEXTILESTEXTILEST

DESIGNSFORTHEHANDICAPPEDDESIGNSFORTHEHANDICAPPEDDESIGNSFORTHEHANDICAPPEDDESIGNSFORTHEHANDICAPPEDDESIGNSFORTHEHANDICAP
FORTHEHANDICAPPEDDESIGNSFORTHEHANDICAPPEDDESIGNSFORTHEHANDICAPPEDDESIGNSFORTHEHANDICAPPEDDESIGNSFORTHEHANDICAPPEDDESI

PBC International, Inc., New York, New York

Distributors to the trade in the United States:
Robert Silver Associates
95 Madison Avenue
New York, NY 10016

Distributors to the trade in Canada:
General Publishing Co. Ltd.
30 Lesmill Road
Don Mills, Ontario, Canada M3B 2T6

Distributed in Continental Europe by:
Feffer and Simons, B.V.
170 Rijnkade
Weesp, Netherlands

Distributed throughout the rest of the world by:
Fleetbooks, S.A.
℅ Feffer and Simons, Inc.
100 Park Avenue
New York, NY 10017

Library of Congress Cataloging in Publication Data

Main entry under title:

Product design.

Includes indexes.
1. Design, Industrial. I. Busch, Aki. II. Industrial design magazine.
TS171.P73 1984 745.2 84-5915
ISBN 0-86636-002-6

PRINTED IN HONG KONG
10 9 8 7 6 5 4 3 2 1

Product Design

publisher: Herb Taylor
project director: Cora S. Taylor
managing editor: Linda Weinraub
managing editor: Steven Holt
ID magazine
editor: Carol Denby
art director: Richard Liu
art associate: Charlene Sison

Typesetting by **Trufont Typographers, Inc.**
Hicksville, New York

Color separation, printing, and binding by
Toppan Printing Co. (H.K.) Ltd., Hong Kong

Contents

Introduction 8

CHAPTER 1
Appliances, Housewares, and Tools 10

CHAPTER 2
Home Electronics and Entertainment 48

CHAPTER 3
Lighting 62

CHAPTER 4
Contract and Residential Furnishings 88

CHAPTER 5
Business Equipment 152

CHAPTER 6
Medical Equipment 172

CHAPTER 7
Industrial Equipment and Transportation 186

CHAPTER 8
Recreational and Sports Equipment 210

CHAPTER 9
Textiles 222

CHAPTER 10
Designs for the Handicapped 236

APPENDIX

Design Awards Programs 246

INDEX I
Products 248

INDEX II
Designers 252

INDEX III
Clients 255

APPLIANCESHOUSEWARESTOOLSAPPLIANCESHOUSEWARESTOOLSAPPLIANCESHOUSEWARESTOOLSAPPLIANCESHOUSEWARESTOOLSAPPLIANCESHOUSEWARESTOOLSAPPLIAN
APPLIANCESHOUSEWARESTOOLSAPPLIANCESHOUSEWARESTOOLSAPPLIANCESHOUSEWARESTOOLSAPPLIANCESHOUSEWARESTOOLSAPPLIANCESHOUSEWARESTOOLSAPPLIAN

HOMEELECTRONICSENTERTAINMENTHOMEELECTRONICSENTERTAINMENTHOMEELECTRONICSENTERTAINMENTHOMEELECTRONICSHOMEELECTRONICSENTERTA
ENTERTAINMENTHOMEELECTRONICSENTERTAINMENTHOMEELECTRONICSENTERTAINMENTHOMEELECTRONICSENTERTAINMENTHOENTERTAINMENTHOMEELECT

LIGHTINGLIGHTINGLIGHTINGLIGHTINGLIGHTINGLIGHTINGLIGHTINGLIGHTINGLIGHTINGLIGHTINGLIGHTINGLIGHTINGLIGHTINGLIGHTINGLIGHTINGLIGHTINGLIG
LIGHTINGLIGHTINGLIGHTINGLIGHTINGLIGHTINGLIGHTINGLIGHTINGLIGHTINGLIGHTINGLIGHTINGLIGHTINGLIGHTINGLIGHTINGLIGHTINGLIGHTINGLIGHTINGLIG

CONTRACTRESIDENTIALFURNISHINGSCONTRACTRESIDENTIALFURNISHINGSCONTRACTRESIDENTIALFURNISHINGS
RESIDENTIALFURNISHINGSCONTRACTRESIDENTIALFURNISHINGSCONTRACTRESIDENTIALFURNISHINGSCONTRACT

BUSINESSEQUIPMENTBUSINESSEQUIPMENTBUSINESSEQUIPMENTBUSINESSEQUIPMENTBUSINESS
BUSINESSEQUIPMENTBUSINESSEQUIPMENTBUSINESSEQUIPMENTBUSINESSEQUIPMENTBUSINESS

MEDICALEQUIPMENTMEDICALEQUIPMENTMEDICALEQUIPMENTMEDICALEQUIPME
EQUIPMENTMEDICALEQUIPMENTMEDICALEQUIPMENTMEDICALEQUIPMENTMEDIC

INDUSTRIALEQUIPMENTTRANSPORTATIONINDUSTRIALEQUIP
INDUSTRIALEQUIPMENTTRANSPORTATIONINDUSTRIALEQUIP

RECREATIONALSPORTSEQUIPMENTRECREAT
EQUIPMENTRECREATIONALSPORTSEQUIPME

TEXTILESTEXTILESTEXT
TEXTILESTEXTILESTEXT

DESIGNS
FORTHEH

Introduction

What we work on is going to be ridden in, sat upon, looked at, talked into, activated, operated, or in some way used by people. . . . If the point of contact between the product and the people becomes a point of friction, then the industrial designer has failed. If people are made safer, more comfortable, more eager to purchase, more efficient—or just plain happier—the designer has succeeded.

—Henry Dreyfuss
Designing for People

The objective of *Product Design* has been to document outstanding designs introduced in the early 1980s. It is not encyclopedic, for there is no attempt to survey specific product categories. The collection, however, *is* comprehensive; it is composed of designs that have received recognition in the international design community.

In approaching the formidable task of gathering material, we sought independent endorsement to help validate products for publication. With such a wide array of forms and functions, personal taste alone could not be authoritative. To begin with, we gathered products that had been recognized by recent international design competitions; these included the *ID Annual Design Review*, the Braun Prize, the ICSID/CID Philips Awards, Britain's Design Council Citations, the Osaka International Design Competition, and the Industrial Design Excellence Awards sponsored by the Industrial Designers Society of America. We also reviewed products that were selected for exhibition by galleries or museums; finally, the editors of *ID*, who are able to screen many new products, worked with Akiko Busch to select additional pieces by consensus.

The question of merit, of course, is always difficult; even when a jury is involved, subjective influences—such as aesthetics, fashion, taste, and the reputation of the designer, manufacturer, or retailer—come into play. Criteria vary, even within categories. In some cases, a product represents such an advanced and wonderful design solution that to include it seems absolutely right. In other instances, product differentiation is subtle and the reasons for inclusion obscure. Such disparate conditions are inevitable in any large collection.

In her lively and studious chapter introductions, Busch sets the stage for each of the ten chapters with relevant examples and an analysis of product design trends. In a collection as large as this one, however, we found it impossible to embark on a case-study approach, and to report the constraints and rationale that apply to each design.

One does not usually find careerist ambitions encapsulated in, say, a tape deck, but a designer *can* identify himself in the work; though there is no signature, the memory of the process can remain forever rich. In this sense, at the core of this collection is a tribute to the design process. As we recognize the elegant design of man-made objects, we recognize individual or collective imagination, intelligence, skill and talent.

What Ettore Sottsass, Jr., the Italian designer, said about a book on his work can be applied to this collection:

"A book is not real life: the daily grind, the anxiety, the confusion, the excuses of a headache or the radio that would not let you work in peace, the thousands of cards covered with sketches that seemed so brilliant because you never risked finishing them, and all those other distractions . . . in a finished book these things are no longer apparent."

The contemporary designer is the centerpiece of this collection, even where the design is anonymous. Though there are inevitable omissions, woven through the ten chapters is something of a who's who in contemporary international design.

The history of design since World War II reflects the uneasy coexistence of two distinct visions of the future, explains Jeffrey Meikle, author of *Twentieth Century Limited*. Both visions are illustrated in this collection; one, self-consciously elitist, stresses the moral, even the spiritual obligation of the designer, the other, more democratic, concentrates on providing the public with what it seems to want. In spite of this tension, designers are frequently able to please the critics *and* the public. Even in this age of perpetual novelty, the level of mass design has improved; at the same time, *haut* design coexists with the Bauhaus ideal that "good" design can help reform society. In the everyday world, it may still be Raymond Loewy's MAYA axiom that is the most practical guide: it suggests that a designer should offer an inherently conservative public the "most advanced yet acceptable" version of a product.

Still another tension exists. There is the financial urge, based on economy of scale, to make products for the global market and the creative urge to produce for idiosyncratic segments that may or may not cross national boundaries. In the last ten years, product design has branched in many directions, pulled to and fro by competing artistic, financial, and marketing philosophies; the seamless modern aesthetic that guided design from the forties to the early seventies has been challenged by a healthy new diversity.

Design excellence does make a tangible contribution to a nation's economic well-being, according to author and economist John Kenneth Galbraith, the Paul Welberg Professor of Economics Emeritus at Harvard University.

". . . design depends not alone on the availability of artists, it invokes the whole depth and quality of the artistic tradition. It is on this that modern industrial success has come largely to depend. Proof is wonderfully evident. . . . One of the miracles of modern industrial development—of modern industrial achievement—has been modern Italy. Since World War II Italy has gone from one major public disaster to another with one of the highest rates of economic growth of any country in the Western industrial world. . . . Italy has been an economic success over the last 35 years because Italian design is better; because its products appeal more deeply to the artistic sense. An Italian design reflects . . . the superb commitment of Italy to artistic excellence extending over the centuries and continuing down to the present day."

There is also an evangelical side to *Product Design* in the sense that we want to help raise the level of design literacy. Though different from the way financial or marketing people think, the "designerly" way of thinking and communicating is vivid, and, in a world where old industrial patterns are fading, an approach to problem-solving that is increasingly relevant. Their combination of intuitive and logical thinking leads to true innovations, authentic product benefits, and elegant forms.

The traditional conflict between expedient marketing goals and design fidelity, has a powerful effect on the finished product. For example, a designer can be told, "Take out ten percent of the cost, but keep the perceived value." The compromise results in what is advertised as "style" and euphuistically called "cost-efficient design." In this common scenario, the designer tries to negotiate trade-offs in the consumer's favor, for it is the designer who is the custodian of the public's interest; in the vernacular, he represents the "end user."

The modern designer is expected to be familiar with aesthetics, engineering, ergonomics, fabrication, materials, marketing and sales. But in the practical world, success depends on team skills. When collaboration is sincere, it is exhilarating and leads to a high degree of design fidelity.

Yet design is never really anonymous, for refined products are made by companies that recognize aesthetic sensibilities, in short, the creativity of the individual. This is true whether the designer works for the corporation or is a member of an independent, consulting design firm. Because teamwork is inevitable, a culture of sensitive collaboration nourishes creativity.

—Randolph McAusland
Publisher, *Industrial Design*

CHAPTER **1**

Appliances Housewares and Tools

When industrial design first established itself as a field in the early days of this century, the products that emerged from the movement often resembled something else entirely. The consumer, in those days, may not have been entirely prepared for lamps, ice boxes, and radios that looked as though they did what they did. In fact, he or she generally preferred forms which, drawn from nature, were familiar. Witness, then, the streamlined iceboxes that appeared to be designed to travel at high speeds, lamps in the shape of dragonflies, and sewing machines gracefully lounging upon beds of ailanthus leaves.

All this, of course, changed with the Modern Movement, which measured the integrity of a form by the degree to which it expressed its function. For decades now, if any generalization could be made about home appliances, it is that their design had to clarify—beyond reasonable doubt—their function. Ailanthus leaves clearly obscured the issue of sewing. And, not only must an object look as though it does what it does, it should also simplify itself for the user. The uncluttered form found in most Braun classics perhaps best expresses this maxim.

While other areas of design may be staging minor revolutions that question these clean and classic lines of Modernism—most notably through a greater use of color and ornamentation—consumer appliances remain less playful. Form must continue to communicate function, not because its designers are unimaginative, but because the growing capabilities of consumer appliances make it all the more necessary for the shape of the object, the visibility of its graphics, and placement of controls all to contribute to the simplicity of operation and to make use self-evident. That is, few appliances have been able to afford to be ornamental without sacrificing some degree of clarity.

Still, no sooner are these words spoken than exceptions appear, and it is these exceptions that are one mark of contemporary design. The instant one passes swift judgment upon the overall purity of consumer appliances, products such as the Cyclon vacuum cleaner confront the eye. Described in one instance as "a successful mixture of a RCA degree show, Memphis, *Star Wars*, Art Deco, *Alien*, and Centre Pompidou styling motifs," it appears at first an anachronism. Indeed, ornamentalism is running rampant here. It shamelessly recollects the design by free association that marked the much earlier days of industrial design. The difference, though, is that the extravagant form of the machine celebrates, rather than conceals, its technical achievements. The purpose of the ailanthus leaves decorating the sewing machine was to suggest nature and thus diminish its possibly intimidating mechanical supremacy: By making the machine appear more "natural," they tried to make it more accessible, less intimidating. In the 1983 vacuum cleaner, the bizarre serves a different purpose: It glorifies the mechanism and flaunts technical innovation, which in this case is the application of centrifugal force to vacuum cleaner design.

It is this, then, that makes for much of the richness found in the design of many of these tools, appliances, and home products: While some quietly and eloquently express their function in simple, classical form, others, in their exuberance, are more likely to shout it.

But contemporary home appliances are distinguished by other features as well. In *A History of Industrial Design*, Edward Lucie Smith points out that the late nineteenth century kitchen could be regarded as a small factory that transformed raw materials into a finished product. While this was certainly true then, it is also true now that the factory has become much smaller while the equipment intended for it has multiplied. In urban areas, especially, the kitchen is often no more than a narrow counter; meanwhile the array of its appliances and the special fixtures, features, and attachments intended for each are produced at a dizzying rate. So clearly, portability, miniaturization, and the ability of a product to file, fold, stack, stash, and perform any number of other reductive contortions to save space marks design excellence in the 1980s.

Or, given the fact that there still may not be enough space to store some of these items, their aesthetics must often be such that one does not mind seeing them even when they are not in use. An example is the prototype for a sculptural fire extinguisher. While its actual use is more likely to inspire terror, it remains a pleasure to look at.

It should perhaps be mentioned here that this and other prototypes have been presented here along with products that have been manufactured and marketed. The criteria used to judge this selection considered form and material more so than the test of the marketplace. While the latter is not insignificant and surely helps to ascertain the success of a product's design, its use as the single criterion for this selection would only exclude innovative and provocative design solutions that are yet in the conceptual or prototype stage.

Recent product design is distinguished as well simply for the range of categories in which it appears. With due thanks to Braun, we now expect calculators and hairdryers to be attentive to design. But screwdrivers and irons we do not. Industries not previously familiar with design have begun to showcase it. This indicates, perhaps, that competitive design work in product development may decide sales and that design entering new territories is another signpost of the 1980s.

If design in untraditional areas is one signpost, a parallel one is that traditionally "designed" products are being designed more and more frequently by professionals working outside the strict boundaries of "industrial design." The silver teapots, for example, commissioned by Alessi, were designed by architects. Available for anywhere from $10,000 to $30,000, they are not exactly products to encourage serious consumer or manufacturer interest. Why, then, have they been included in a collection of product design? For the simple reason that invention can and often will come from without rather than within. Industrial designers can become so absorbed in the physics of material and process that they can bypass simple solutions. For this reason, it can be helpful to glance at the design solutions found elsewhere—solutions not of the economic use of material or process, for ones of pure form. There is a second reason as well: While these tea services may be prohibitively expensive, difficult to produce, and even more difficult to pour and drink from, they more successfully complete another part of their function. That is, they acknowledge and even celebrate the ritual of drinking tea. It is this less tangible function that can so easily be forgotten by the designer who designs housewares on a more regular and repetitive basis. What can designers of buildings say to designers of tableware? Can there be aesthetic cross references? While this selection does not necessarily try to answer these questions, asking that we consider them will surely only help to broaden our vision. And that it might promote a more collaborative spirit between the two professions justifies it all the more.

It also may be helpful to point out that there are absences in this broad category of products. Certainly, the items covered here have a wider range than those in any other category. Suitcases and wrenches, clocks and teapots are included. Still, freezers and dryers are not. It is not the aim here to represent the best of each appliance, but rather, by showing three well-designed vacuum cleaners, to show some of the ways in which it can be done, and by omitting freezers and dryers, to imply that they are areas that invite the designer's fine tuning.

Product:	Radius Two Collection
Designer:	William Sklaroff Philadelphia, Pennsylvania
Design Firm:	William Sklaroff Design Associates Philadelphia, Pennsylvania
Client:	Smith Metal Arts Buffalo, New York
Awards:	1981 IBD Product Design honorable mention
Materials:	In mirror brass, antique brass, mirror aluminum, mirror bronze, statuary bronze, or mirror black

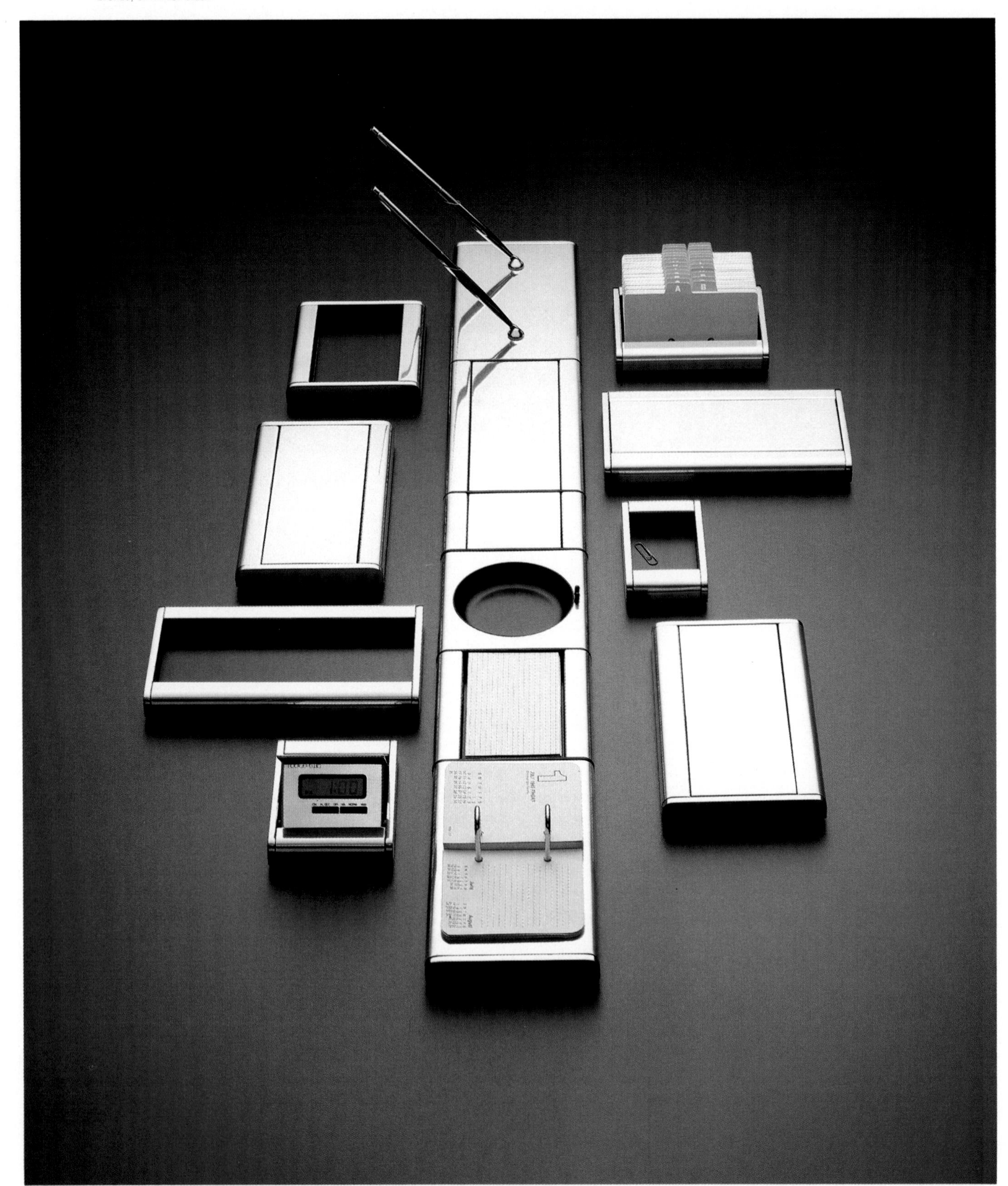

Product:	Safari Pens: Ball Point Pen, Mechanical Pencil, Inkwriter, and Fountain Pen
Designer:	Wolfgang Fabian Heidelberg, West Germany
Client:	C. Josef Lamy GmBH Heidelberg, West Germany
Materials:	ABS plastic

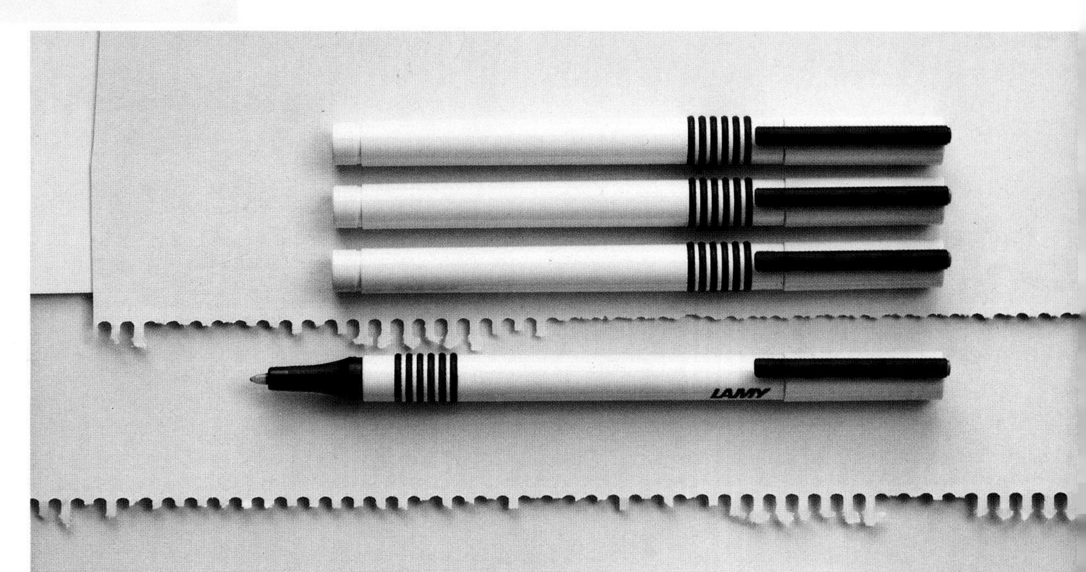

Product:	Lamy White Pens
Designer:	Wolfgang Fabian Heidelberg, West Germany
Client:	C. Josef Lamy GmBH Heidelberg, West Germany
Awards:	Stuttgart Design Center award Hanover Gute Industrie Firm
Materials:	White Makrolon casing, black polypropylene rings, black Delrin clip, and black Novodur cone

Product:	Desk Accessories
Designers:	Toru Nishioka and Studio 80 Tokyo, Japan
Client:	Courtesy Gallery 91 New York, New York
Materials:	Aluminum

Product: Maya
Designer: Tias Eckhoff
Bergen, Norway
Client: Norsk Stalpress A/S
Bergen, Norway
Materials: Stainless steel

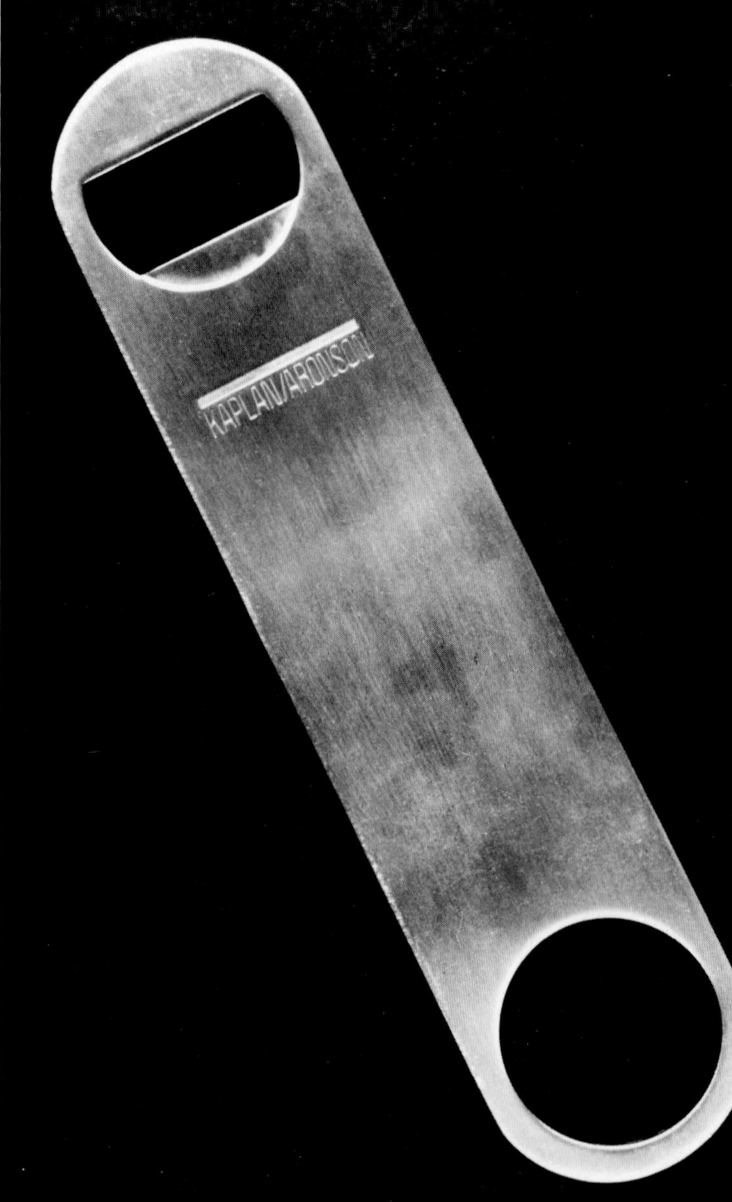

Product: Bottle Opener
Designer: Henry Altchek
New York, New York
Client: Kaplan/Aronson
New York, New York
Awards: 1981 *Industrial Design* magazine Design Review selection
Materials: Stainless steel with brushed satin finish

Product: Century flatware
Designers: Tapio Wirkkala and K. G. Hansen
Client: Rosenthal Studio-Line, Rosenthal AG
Selb, West Germany
Awards: 1983 Die gute Industrieform award
Materials: Silver, aluminiumoxid-Ceramics handle, chromium-Molybdene steel blade

Product: Design 10 Plastic Flatware
Designer: Don Wallance
Croton-on-Hudson, New York
Client: H. E. Lauffer Co., Inc.
Somerset, New Jersey
Materials: Injection-molded Lexan (polycarbonate) with matte finish

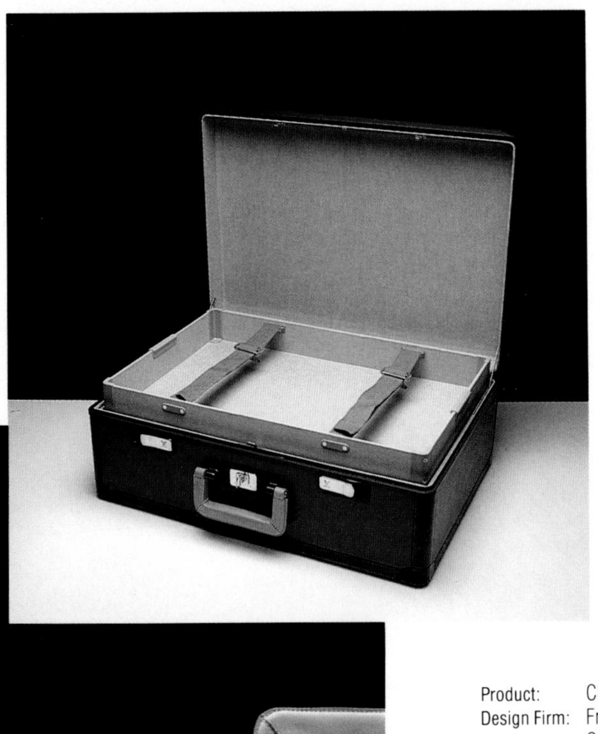

Product:	Challenge Case and Briefcase
Design Firm:	Frogdesign Campbell, California
Client:	Louis Vuitton Paris, France
Materials:	Kevlar/polyurethane canvas; brass or chrome fittings, natural leather

Product:	American Tourister Business Equipment
Designer:	Ed Lawing, Richardson/Smith Worthington, Ohio
Design Firm:	Richardson/Smith Worthington, Ohio
Client:	American Tourister Warren, Rhode Island
Materials:	Injection-molded Lexan frame with vacuum-formed shells wrapped in top-grain cowhide

Product:	Swedish attache case
Courtesy:	Conran's New York, New York
Materials:	Pressed cardboard with black plastic hardware

Product: Essteele "Australis"
Client: Silcraft Sales PTY
Victoria, Australia
Awards: 1983 Australian Design award
Materials: Stainless steel, solid copper base

Product: Casserole
Designer: Richard Nissen
Langaa, Denmark
Design Firm: Richard Nissen, A-S
Langaa, Denmark
Courtesy: Sointu
New York, New York
Materials: Cast iron, beechwood handles

Product:	Teapot with brass whistle sounding E and B notes
Designer:	Richard Sapper Milan, Italy
Client:	Alessi Milan, Italy
Courtesy:	Sointu New York, New York
Materials:	Stainless steel, heat diffusing copper bottom, polyamide covered handle

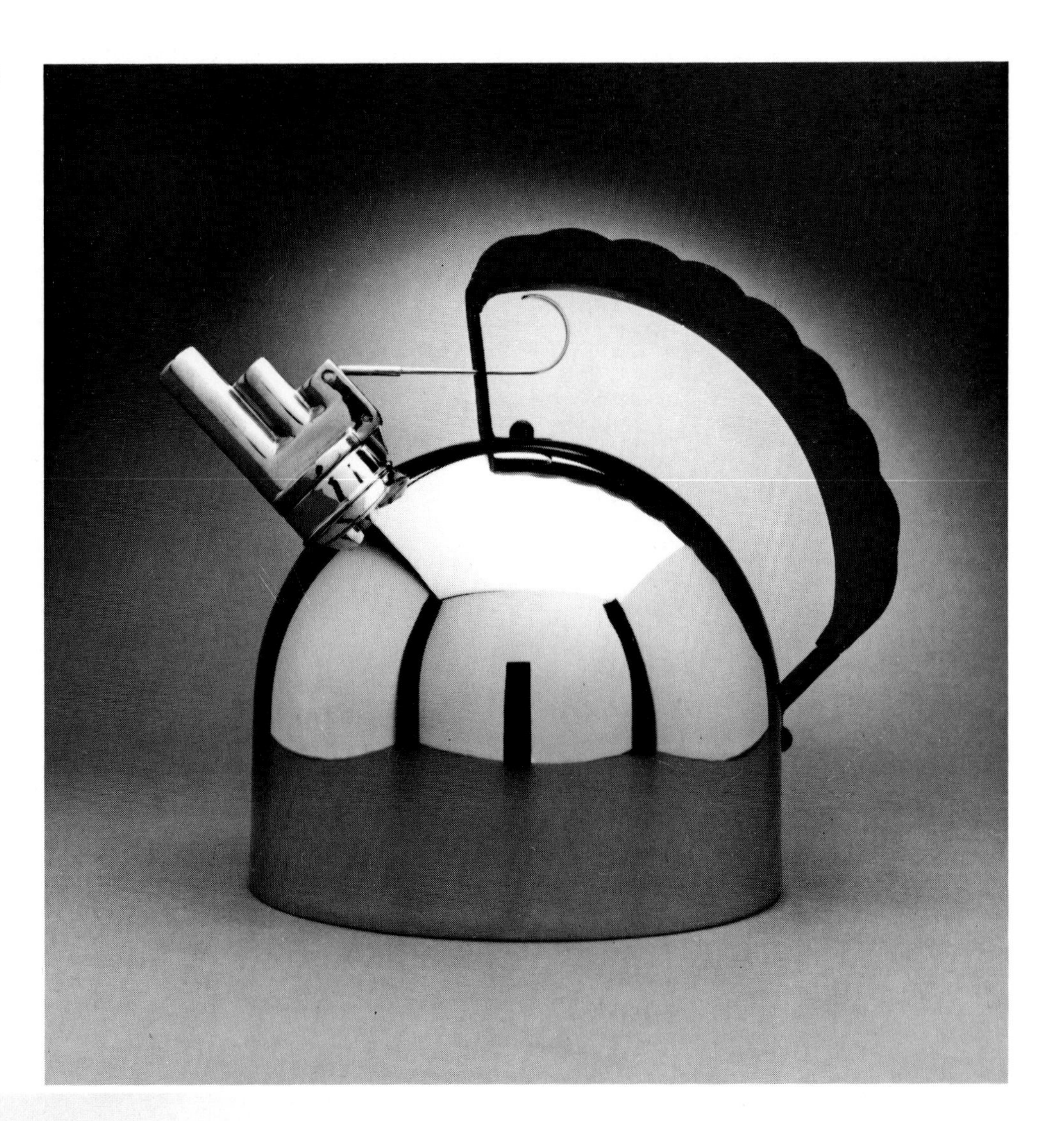

Product:	Bistro tempered teapot and glasses
Courtesy:	Conran's New York, New York
Materials:	Glass, cork trivet, plastic details

Product: Tea and Coffee Service
Designer: Alessandro Mendini
Milan, Italy
Client: Alessi S.p.A
Milan, Italy
Materials: Silver

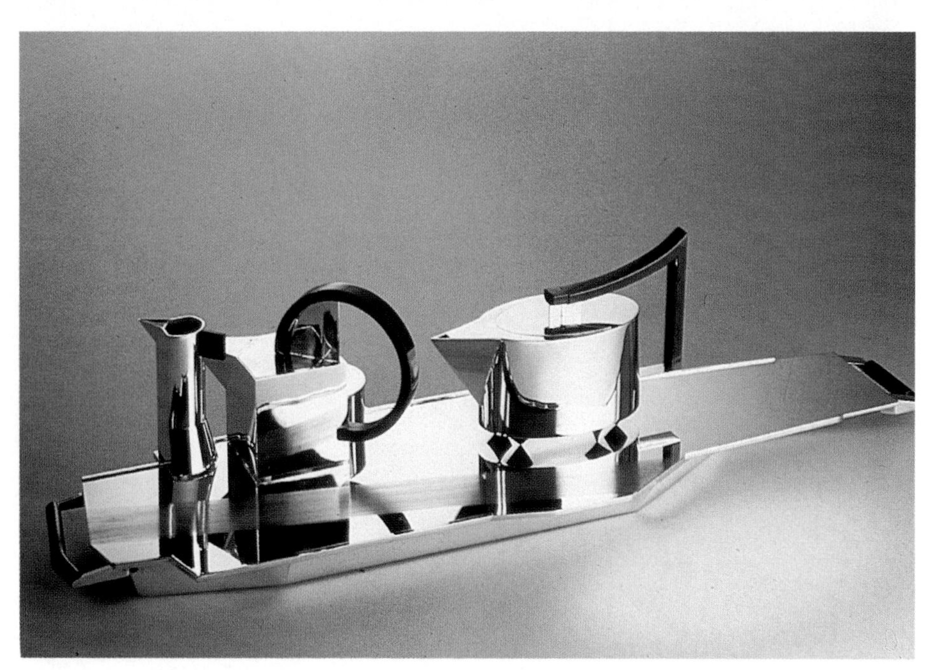

Product: Tea and Coffee Service
Designer: Hans Hollein
Vienna, Austria
Client: Alessi S.p.A.
Milan, Italy
Materials: Silver; knobs, handles, and feet in blue metacrylate

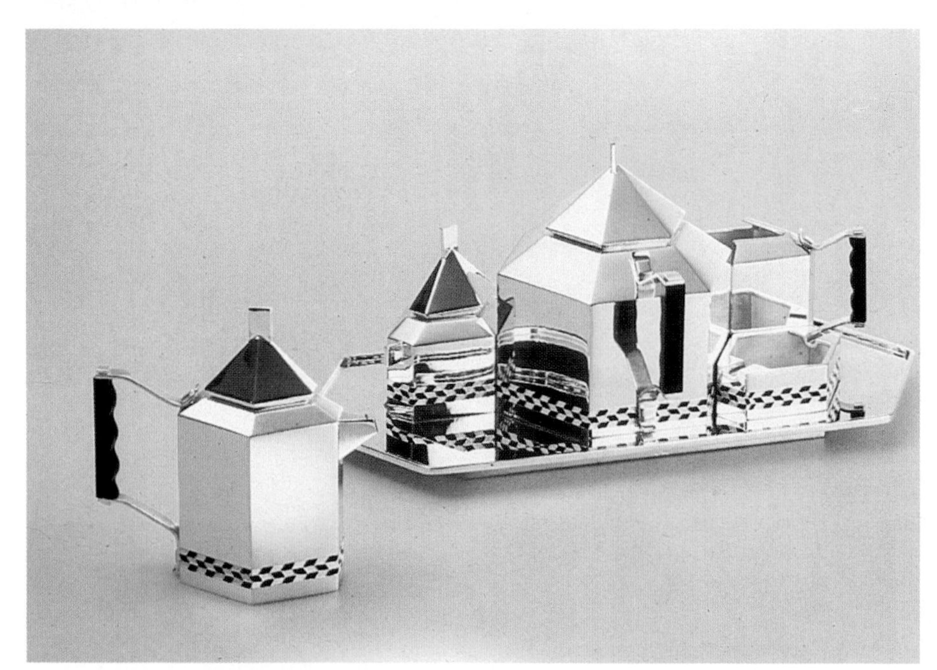

Product: Tea and Coffee Service
Designer: Paolo Portoghesi
Rome, Italy
Client: Alessi S.p.A.
Milan, Italy
Materials: Silver, ebony handles, decorative enamelled black and white bands

Product:	Reticelli
Designer:	Michael Boehm Selb, West Germany
Client:	Rosenthal Studio-Line, Rosenthal AG Selb, West Germany
Materials:	Filigree glass

Product:	Ambassador
Designer:	Stephen Bartlett London, England
Design Firm:	BIB Design Consultants London, England
Client:	Royal Stafford China Limited Stoke-on-Trent, England
Awards:	1982 Design Centre Selection award
Materials:	Bone china

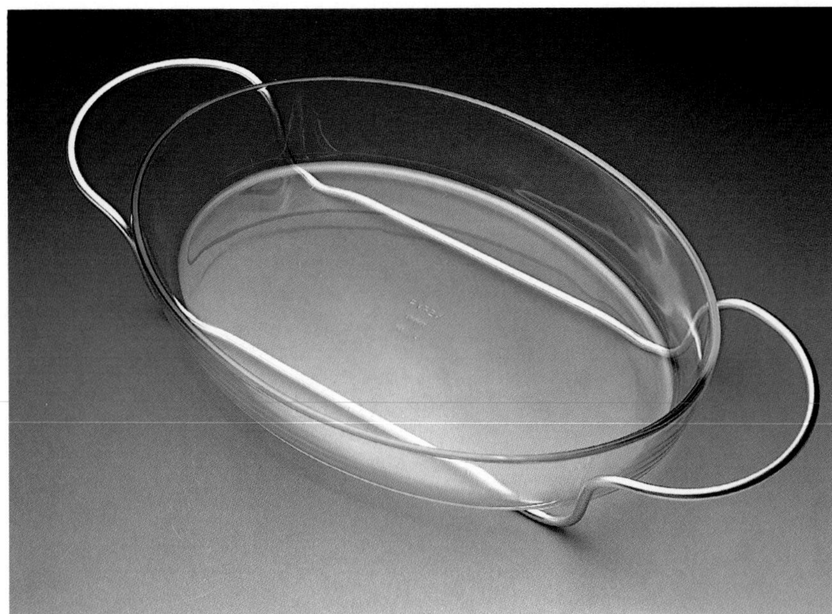

Product: Culinaria
Designers: Davin Stowell and Linda Celentano
New York, New York
Design Firm: Davin Stowell Associates
New York, New York
Client: Corning Designs
Clinton, New Jersey
Materials: Pyrex and plated brass

Product: Marimekko for Pfaltzgraff Collection
Designer: Ristomatti Ratia
Stamford, Connecticut
Client: Pfaltzgraff
York, Pennsylvania
Materials: Stoneware

Product: Fruit Tray
Designer: Masanori Umeda
Tokyo, Japan
Design Firm: Courtesy Gallery 91 New York, New York
Client: Courtesy Gallery 91 New York, New York
Materials: Aluminum

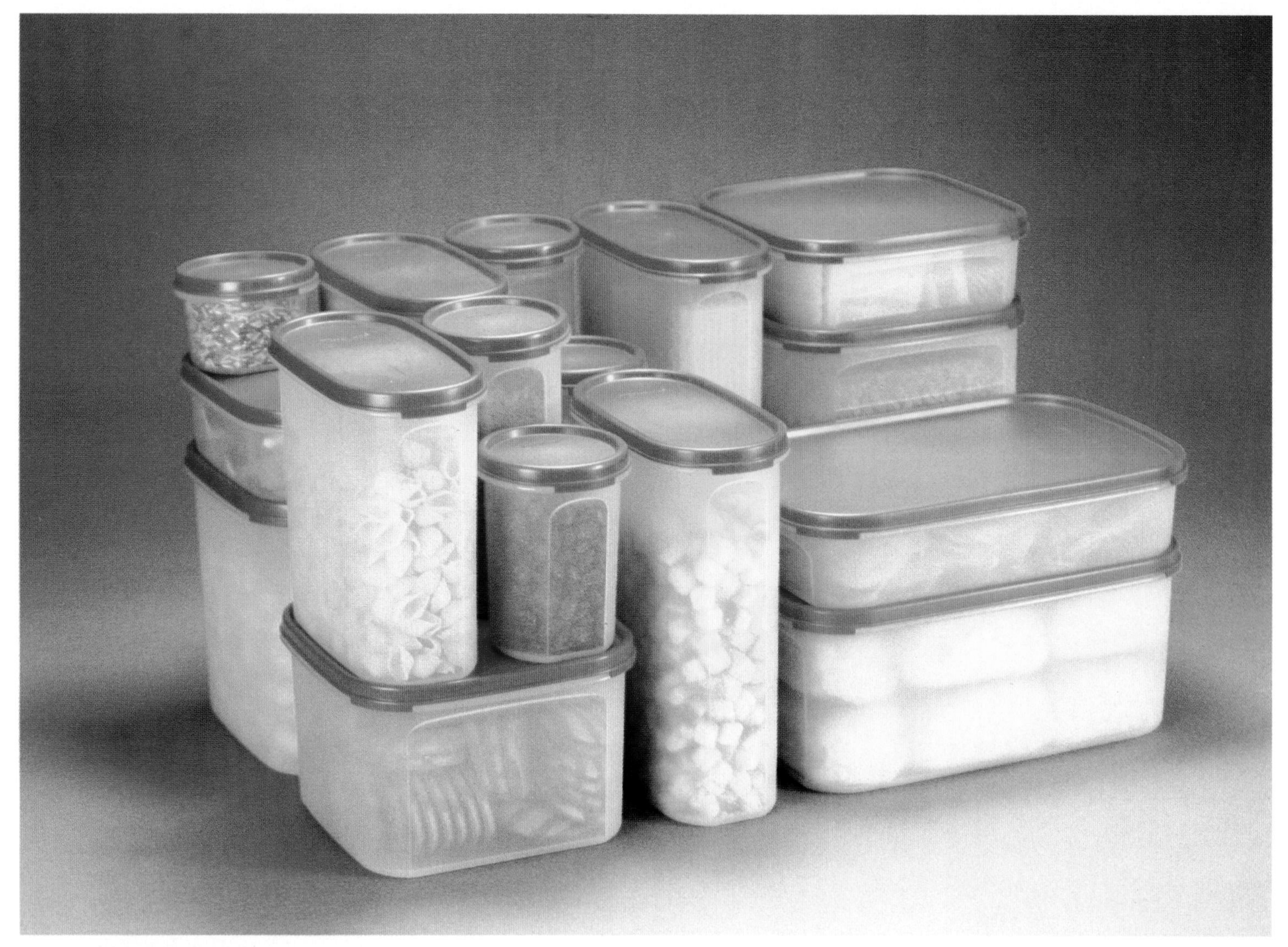

Product:	Tupperware Modular Mates ™ Container System
Designers:	Tupperware Design Groups in U.S.A. and Europe
Client:	Tupperware Woonsocket, Rhode Island
Awards:	1983 *Industrial Design* magazine Design Review selection
Materials:	Polypropylene containers; low density polyethylene seals

Product:	K—Line of cigarette lighter, ash trays, and coasters
Designer:	Masayuki Kurokawa Tokyo Japan
Client:	Fuso Gomu Ind. Co. Ltd Japan
Courtesy:	Sointu New York, New York
Materials:	Cigarette lighter: black styrene; Ashtray: black rubber with stainless steel interior and styrene shell; Coasters: black embossed rubber

Product:	Marimekko for Pfaltzgraff Collection
Designer:	Ristomatti Ratia Stamford, Connecticut
Client:	Pfaltzgraff York, Pennsylvania
Materials:	Stoneware

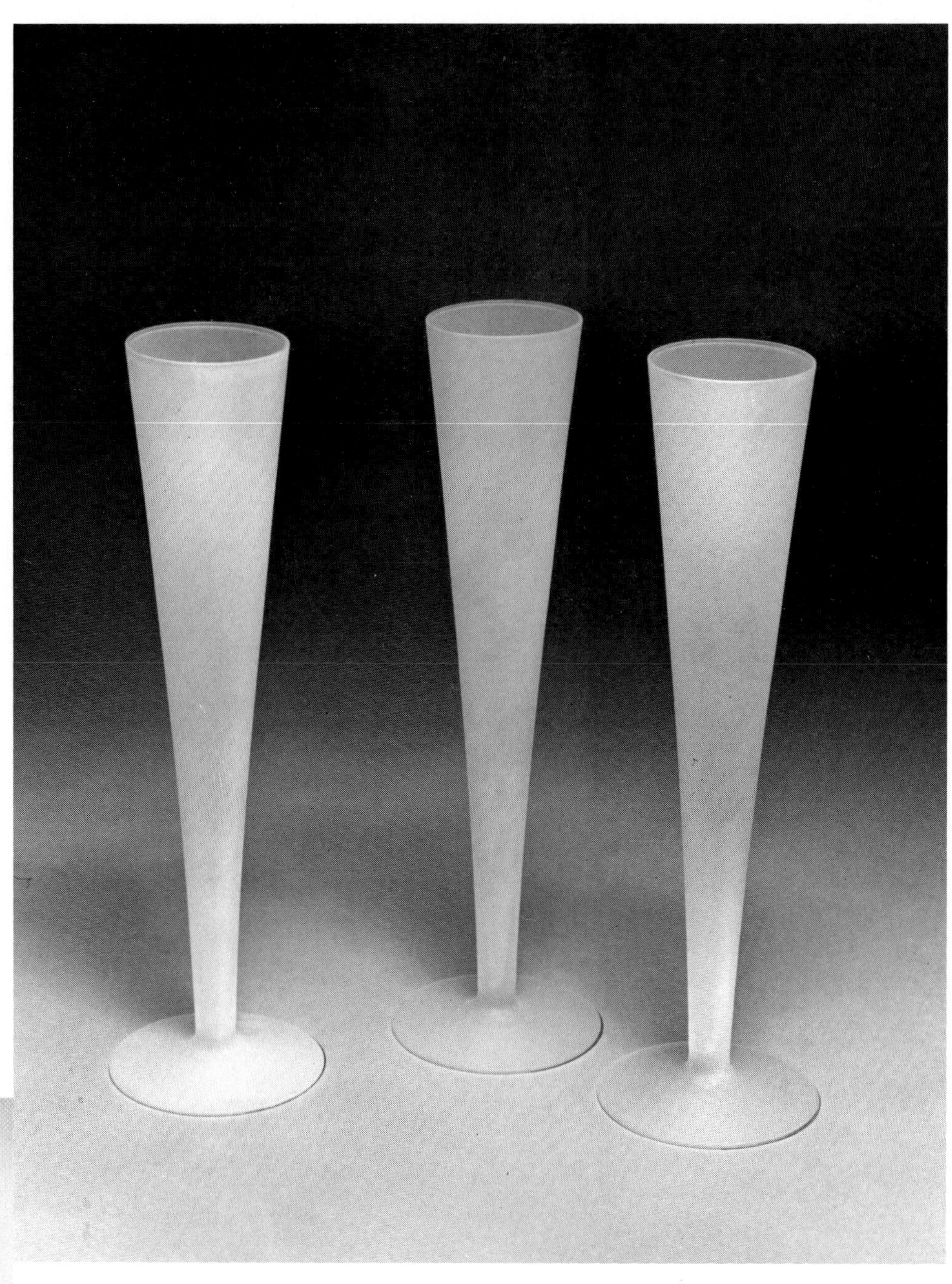

Product: Frosted Champagne Flutes
Manufacturer: Hergiswiler Glaf AG
Hergiswil, Switzerland
Courtesy: Sointu
New York, New York
Materials: Frosted glass

Product: Tea and Coffee Service
Designer: Aldo Rossi
Venice, Italy
Client: Alessi S.p.A.
Milan, Italy
Materials: Silver; blue stoved enamel bands;
Storage unit: black iron base, glass walls, brass frame.

Product: Executive Thermo Pitcher
Courtesy: Conran's
New York, New York
Materials: Plastic with glass thermal liner

Product: Cruet Set
Designer: Ettore Sottsass
Milan, Italy
Client: Alessi
Milan, Italy
Courtesy: Sointu
New York, New York
Materials: Crystal with weighted bottoms; polished stainless steel caps and tray

Product: Juice Jugs and Cups
Designer: Koen de Winter
Beaconsfield, Quebec, Canada
Client: Mepalservice
Lochem, The Netherlands
Awards: Selected for Design Collection of the Museum of Modern Art in New York City
Materials: S.A.N.

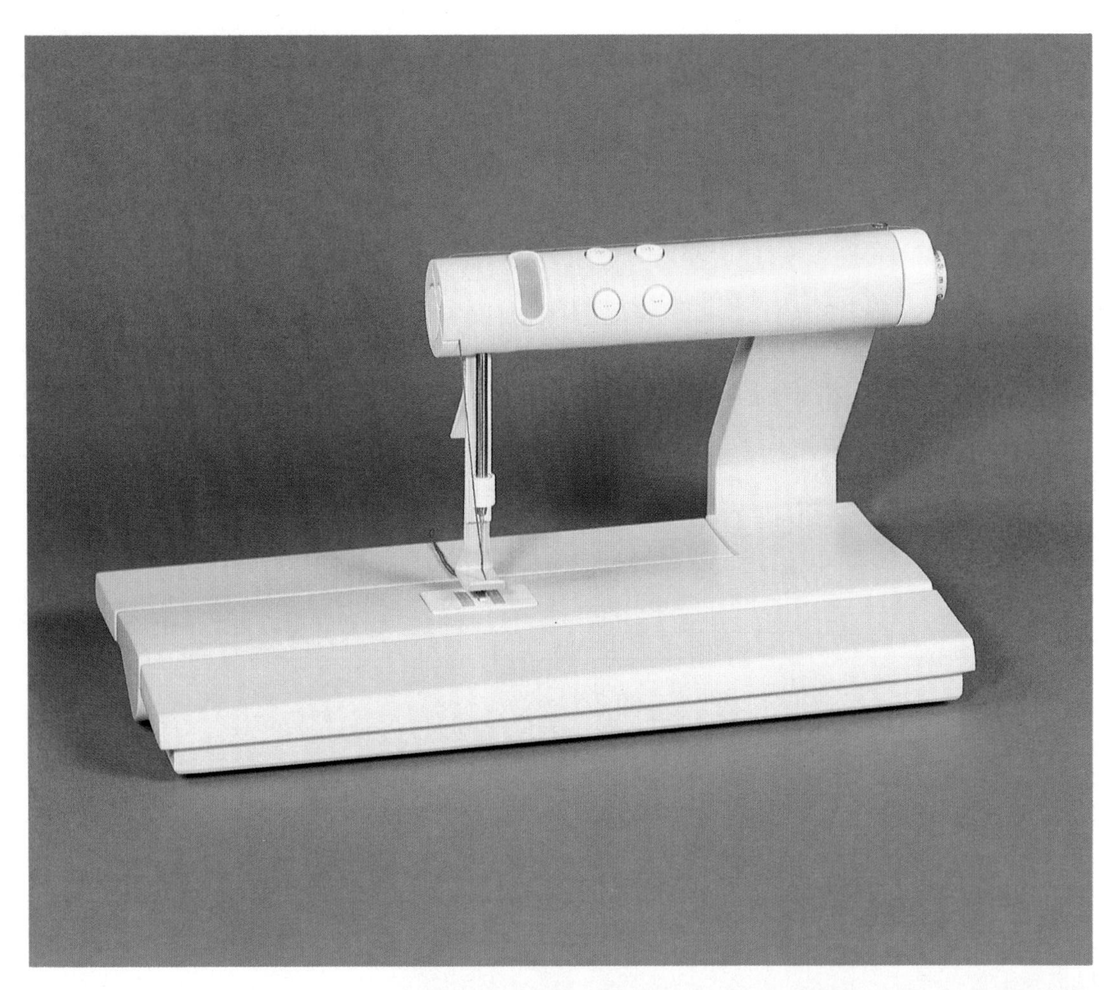

Product: Domestic Sewing Machine
Designer: Roy S. W. Tam
Essex, England
Awards: 1983 Braun Prize for Technical Design

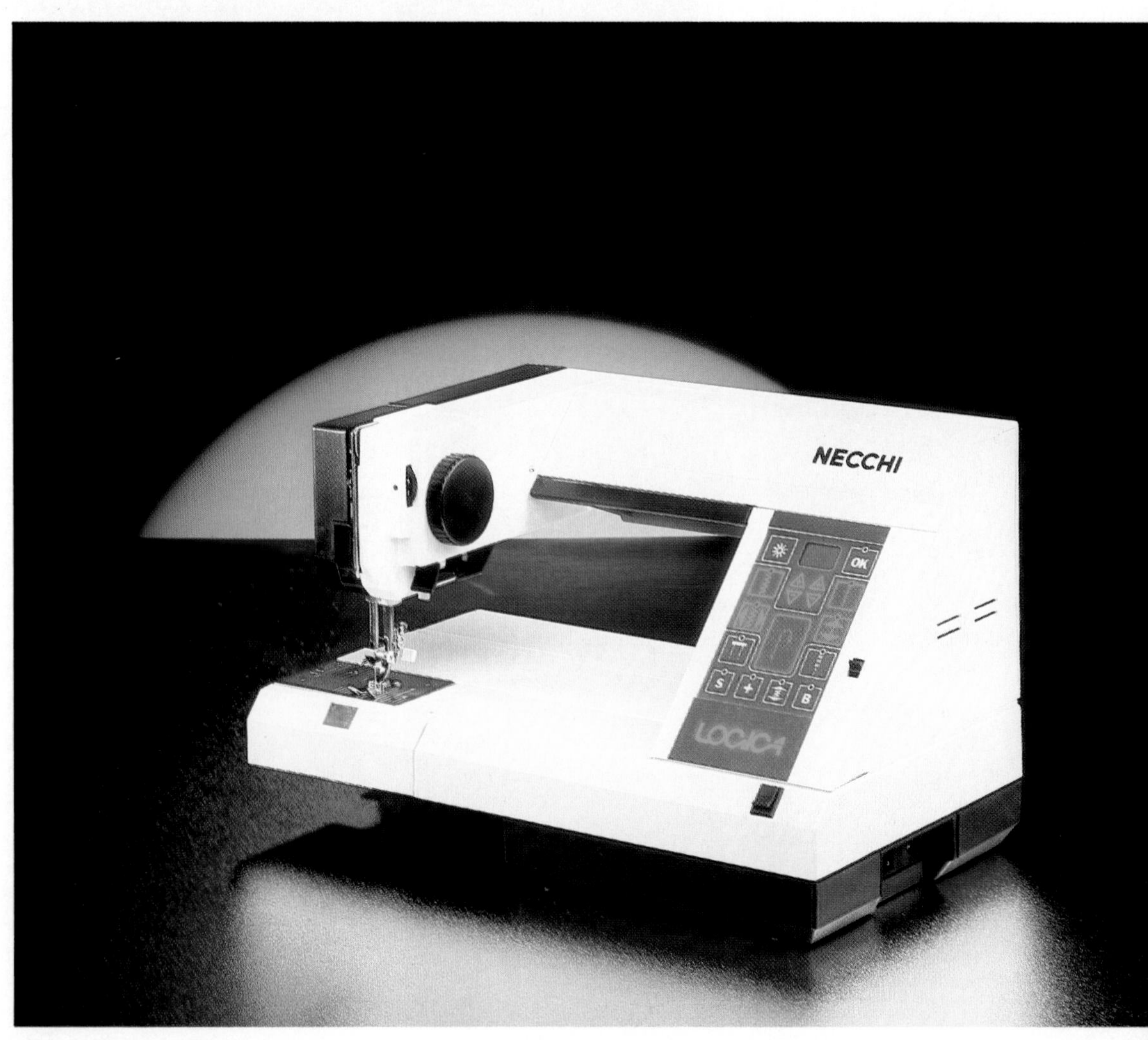

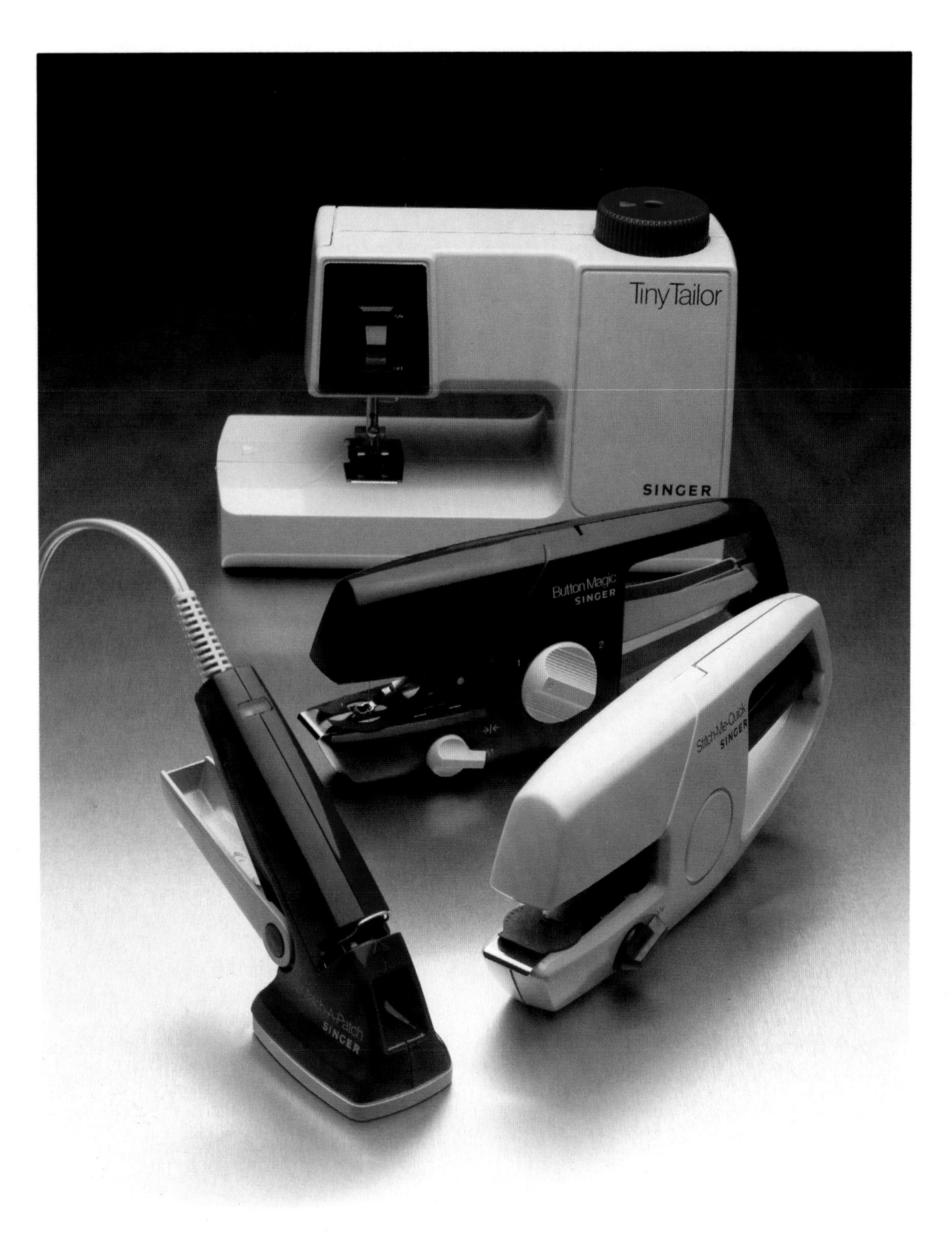

Product: Singer Easy Menders: Tiny Tailor Mending machine, Button Magic Button Sewer, Stitch-Me-Quick Hem and Seam Mender, and Match-A-Patch Hole and Tear Mender
Designers: Michael Laude, Robert Dawson, Thomas Pendleton, and Andrew King Blance Casey Inc.
New Canaan, Connecticut
Design Firm: King Casey Inc.
New Canaan, Connecticut
Client: The Singer Company
Stanford,Connecticut
Materials: Tiny Tailor; glass-filled thermo-set plastic; Button Magic Button Sewer and Stitch-Me-Quick Hem and Seam Mender: styrene and ABS; Match-A-Patch Hole and Tear Mender: ryton and polycarbonate

Product: Logica Electronic Sewing Machine
Designer: Giorgetto Giugiaro
Moncalieri, Torino, Italy
Design Firm: Italdesign
Moncalieri, Torino, Italy
Client: Necchi
Pavia, Italy
Materials: Die-cast aluminum alloy, ephossidic resin electrostatic painted

Product: CHRONOGYR room thermostat for small heating systems
Design Firm: Fellmann Design AG
Dietlikon/ZH, Switzerland
Client: LGZ Landis & Gyr Zug AG
Zug, Switzerland
Materials: Electronic circuit housed in gray-brown plastic

Product: Flip-top Durabeam
Design Firm: BIB Design Consultants
London, England
Client: Duracell UK
West Sussex, England
Materials: Black ABS plastic, yellow acetal moldings

Product: Home Fire Extinguisher
Designer: Gary C. Johnson
Ferndale, Michigan
Materials: ABS plastic shroud, nylon reinforced plastic nozzle and attachment, fiberglass reinforced plastic or metal tank, matte finish with textured grip

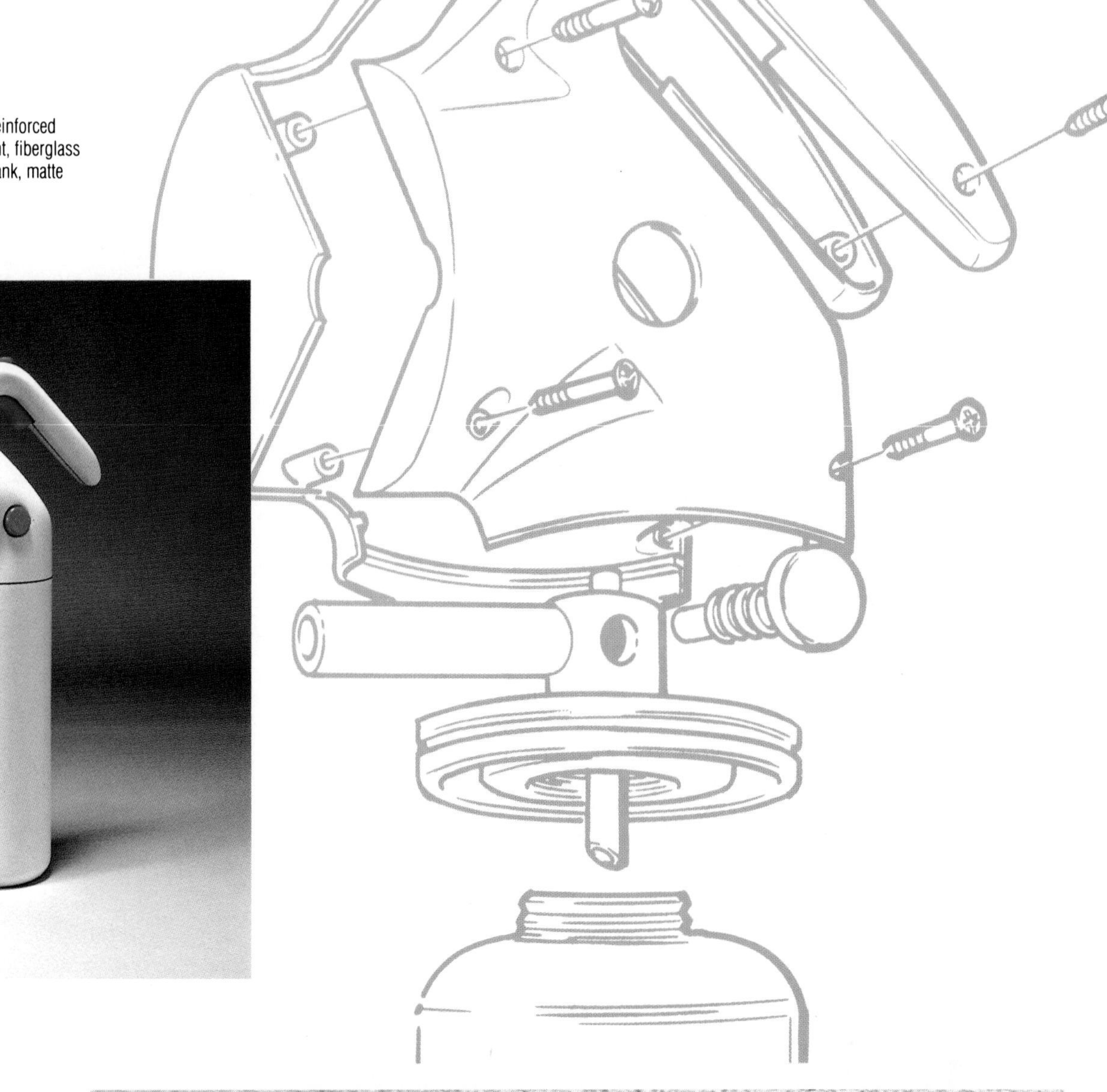

Product: Fan Heater
Designer: Dieter Rams, Braun AG
Kronberg, West Germany
Client: Braun AG
Kronberg, West Germany

Product: Allegroh
Design Firm: Frogdesign
Campbell, California
Client: Hansgrohe
West Germany
Awards: Stuttgart Design Center Excellence of Design award
Materials: Steel ball, teflon covered joint

Product: Magazine Rack
Designer: Ann Maes
Eindoven, The Netherlands
Design Firm: Ann Maes Industrial Design
Eindoven, The Netherlands
Courtesy: Sointu
New York, New York
Materials: Steel with black enamel finish

Product: Mixer Tap
Designer: Vladimir Pezdirc
Ljubljana, Yugoslavia
Design Firm: Studio Kvadrat
Ljubljana, Yugoslavia
Client: Unitas Tovarna Armatur
Ljubljana, Yugoslavia
Materials: Stainless steel

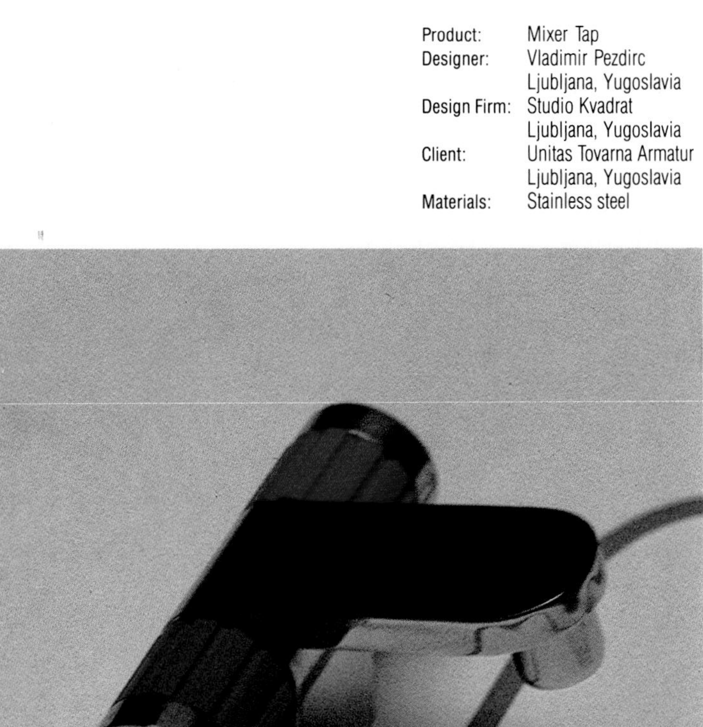

Product: Fireplace Tools
Designer: Ann Maes
Eindoven, The Netherlands
Design Firm: Ann Maes Industrial Design
Eindoven, The Netherlands
Courtesy: Sointu
New York, New York
Materials: Steel with black enamel finish

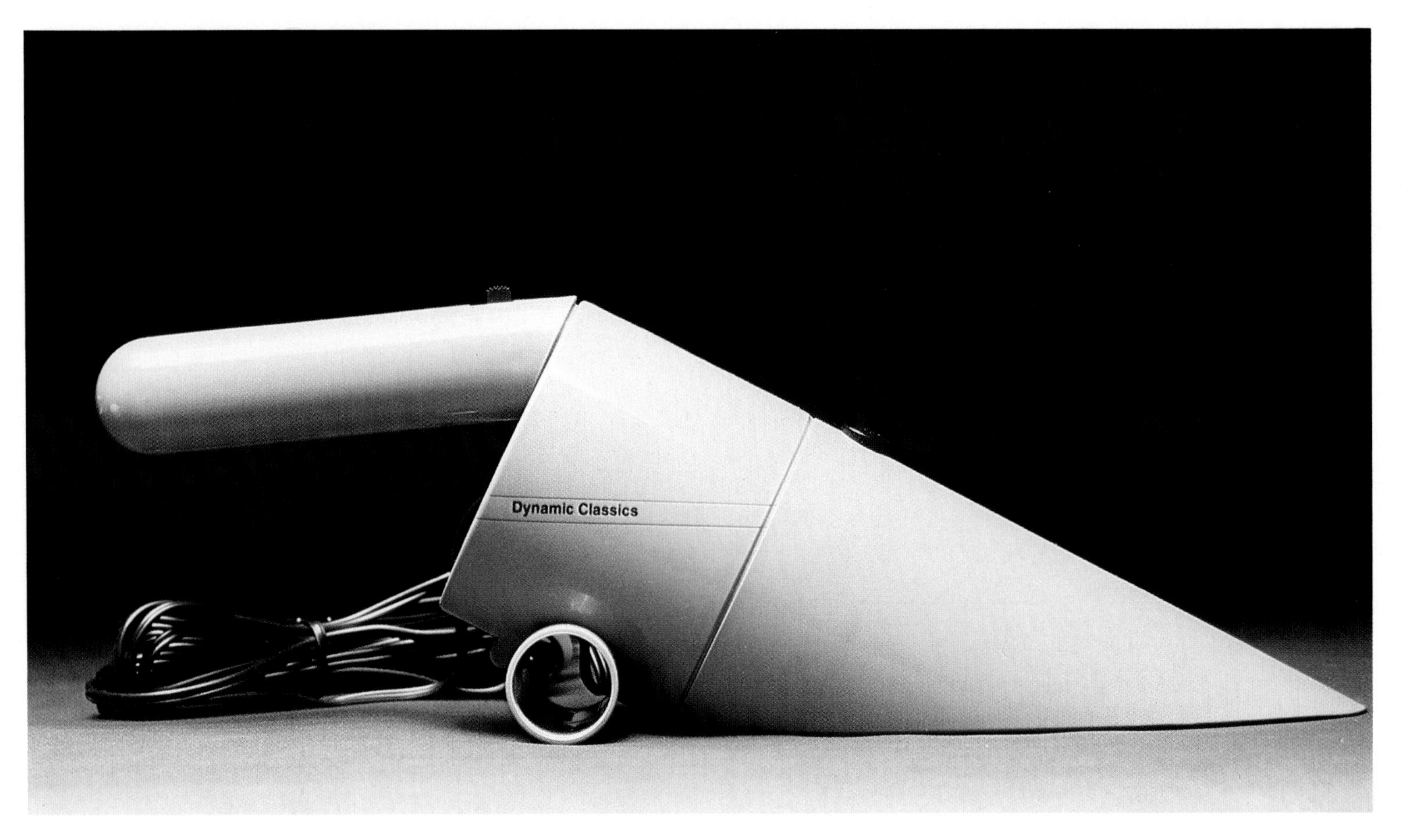

Product:	Dynamic Classics Auto Vacuum
Design Firm:	Morrison S. Cousins and Associates New York, New York
Client:	Dynamic Classics New York, New York
Awards:	1982 *Industrial Design* magazine Design Review selection
Materials:	High-density polyethylene

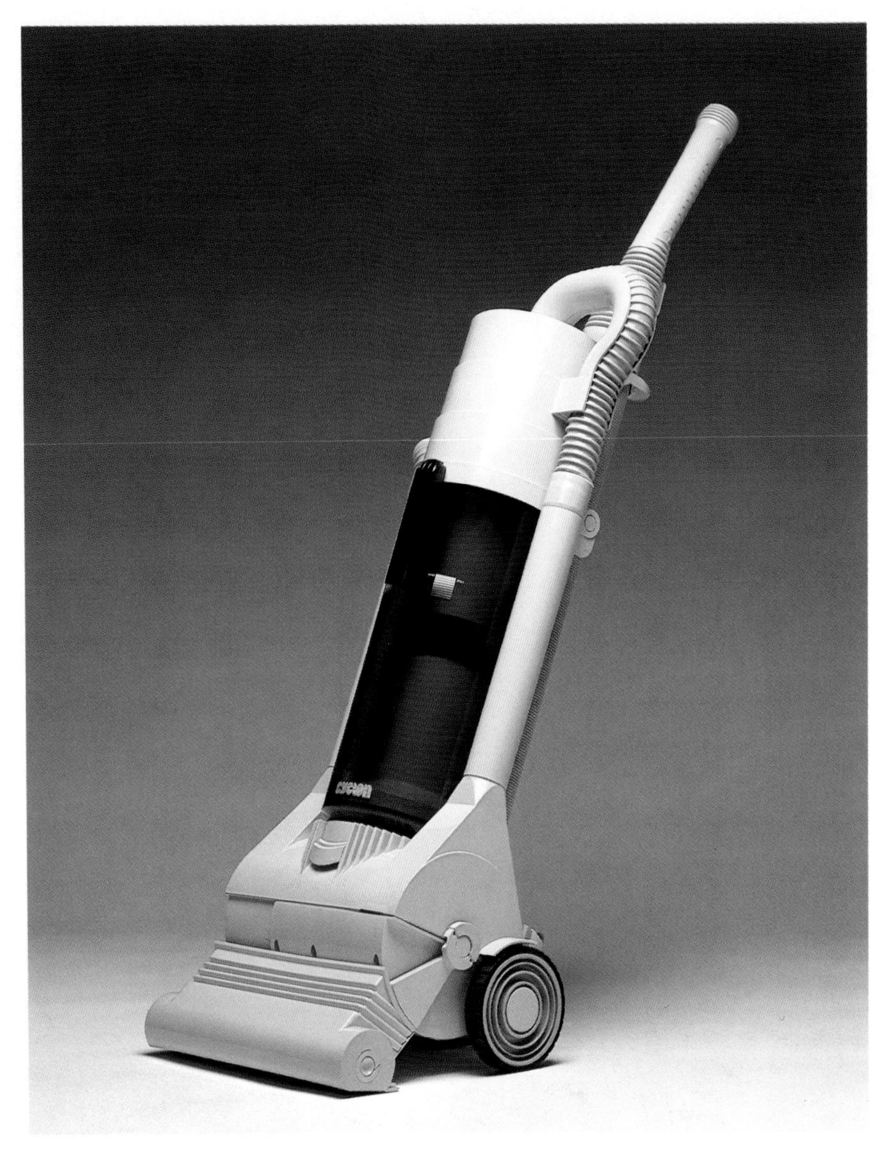

Product: Cyclon Vacuum Cleaner
Designer: James Dyson
Bath, England
Design Firm: Prototypes
Bath, England
Client: Industrie Zanussie S.p.A.
Pordenone, Italy
Materials: ABS body, Macrolon polycarbonate, PVC hose, polyurethane hose cuffs

Product: Eureka Mighty Mite
Designer: Kenneth R. Parker and Samuel E. Hohulin
Bloomington, Illinois
Client: The Eureka Company
Bloomington, Illinois
Materials: Injection molded ABS

Product: Eltron Shavers: Women's Battery Shaver; Model 900; Universal
Designer: Dieter Rams
Kronberg, West Germany
Client: Braun AG
Kronberg, West Germany
Materials: Women's Battery Shaver and Model 900: ABS plastic; Universal: stainless steel case

Product: Airmail Travel Hair Dryer
Designers: Tucker Viemeister, Tamara Thomsen, and John Lonczak
New York, New York
Design Firm: Davin Stowell Associates
New York, New York
Client: Sanyei America Corporation
Fort Lee, New Jersey
Materials: ABS

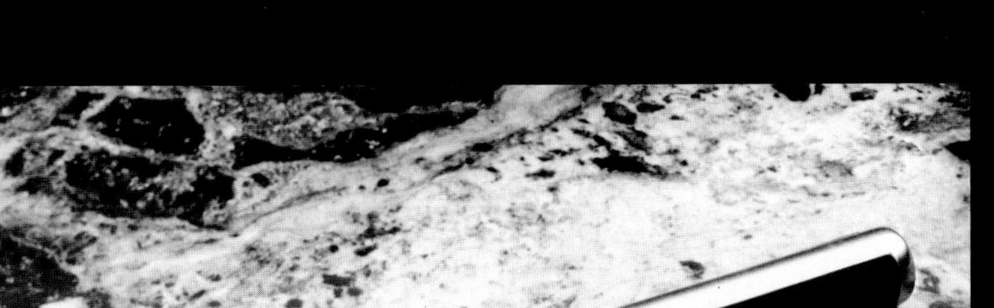

Product: Razor
Designer: Kenneth Grange
London, England
Client: Wilkinson Sword Limited
Buckinghamshire, England
Courtesy: Sointu
New York, New York
Materials: Stainless steel with ribbed black rubber grip

Product:	Serengeti Sunglasses
Designers:	Davin Stowell, Tucker Viemeister, Daniel Formosa, and Tom Dair New York, New York
Design Firm:	Davin Stowell Associates New York, New York
Client:	Corning Glass Works Corning, New York
Materials:	Frames: Plastic with baked epoxy or nickel silver with plated or copper metallic coating. Copper lens color provides sun and glare protection and changes in bright sunlight to rich brown.

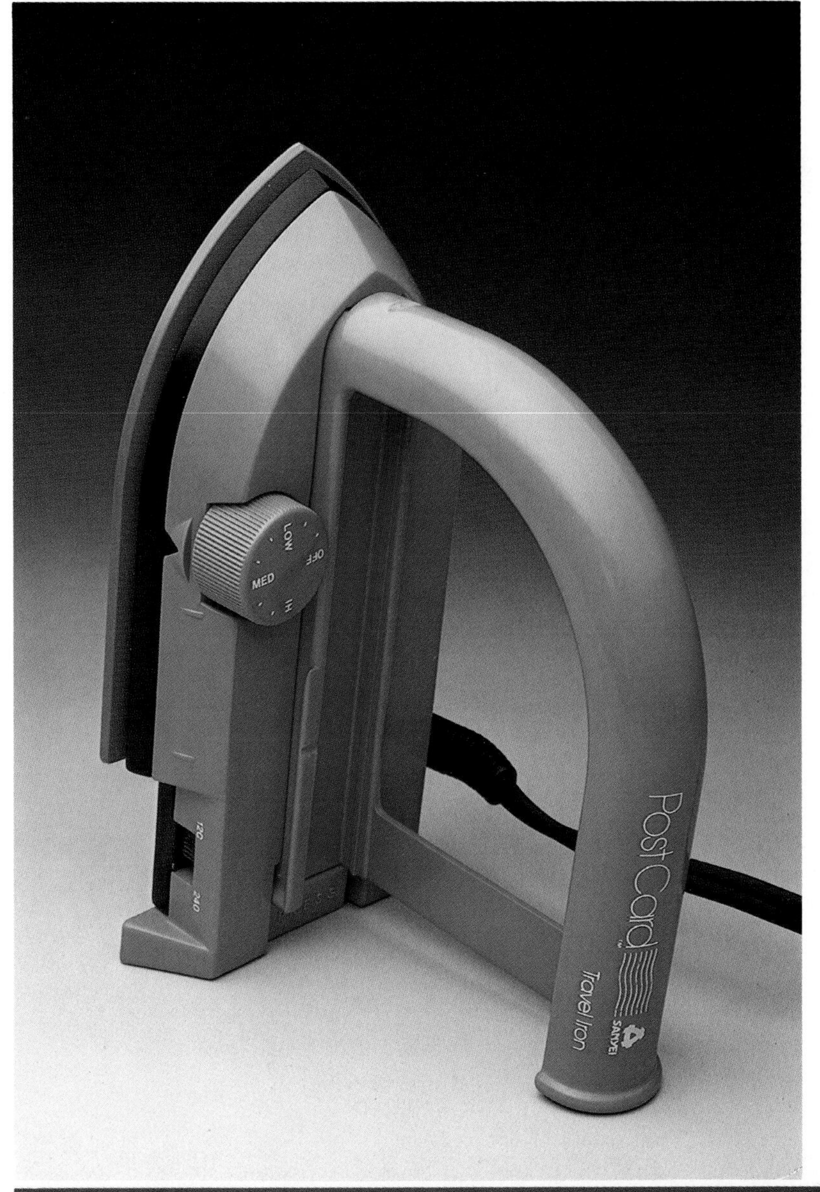

Product: Post Card Travel Iron
Designers: Davin Stowell, Tucker Viemeister, Tom Dair, Tamara Thomsen, John Lonczak, and Daniel Formosa
New York, New York
Design Firm: Davin Stowell Associates
New York, New York
Client: Sanyei America Corporation
Fort Lee, New Jersey
Awards: 1983 *Industrial Design* magazine Design Review selection
Materials: Heat resistant glass-filled polyester, black phenolic layer of insulation, nonstick coating

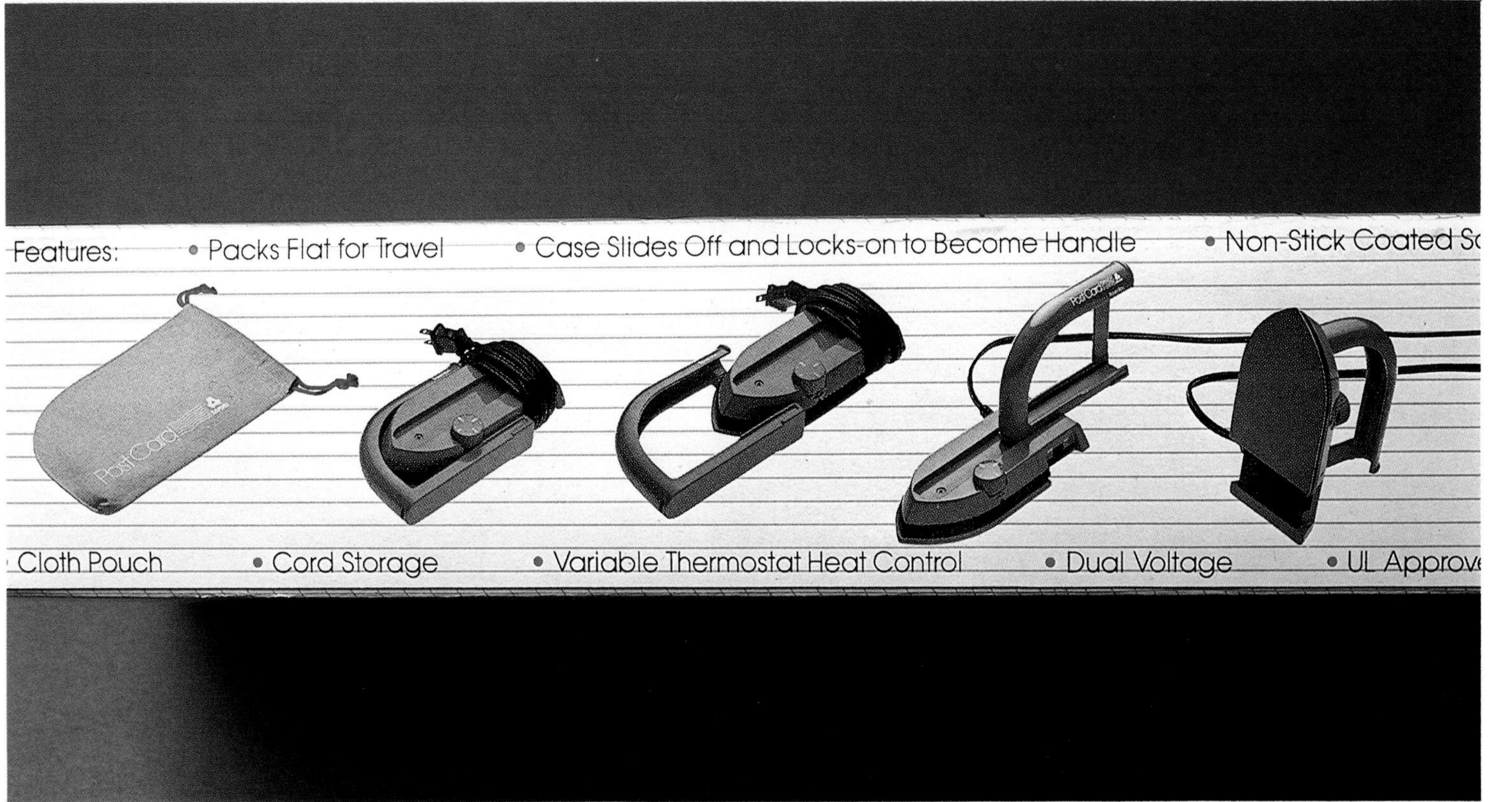

Product: Time of the Earth
Designer: Osamu Akiyama
Awards: 1st International Design Competition
1983
Osaka, Japan

Product: ETA 80 Telephone Set
Designer: Davorin Savnik
Ljubljana, Yugoslavia
Client: ISKRA
Ljubljana, Yugoslavia
Awards: 1981 Stuttgart Design Centre Award
1980 Die Gute Industrieform, Hanover, West Germany
8th Biennial of Industrial Design, Ljubljana, Yugoslavia
Materials: ABS plastic

Product: Vacform Wall Clock
Designer: Stephen Morgan
London, England
Materials: Precolored polystyrene

Product: Braun Calculator
Designer: Dieter Rams
Kronberg, West Germany
Client: Braun AG
Kronberg, West Germany
Materials: ABS plastic

Product: Radius two Collection
Designer: William Sklaroff
Philadelphia, Pennsylvania
Design Firm: William Sklaroff Design Associates
Philadelphia, Pennsylvania
Client: Smith Metal Arts
Buffalo, New York
Awards: 1981 IBD Product Design honorable mention
Materials: In mirror brass, antique brass, mirror aluminum, mirror bronze, statuary bronze, or mirror black

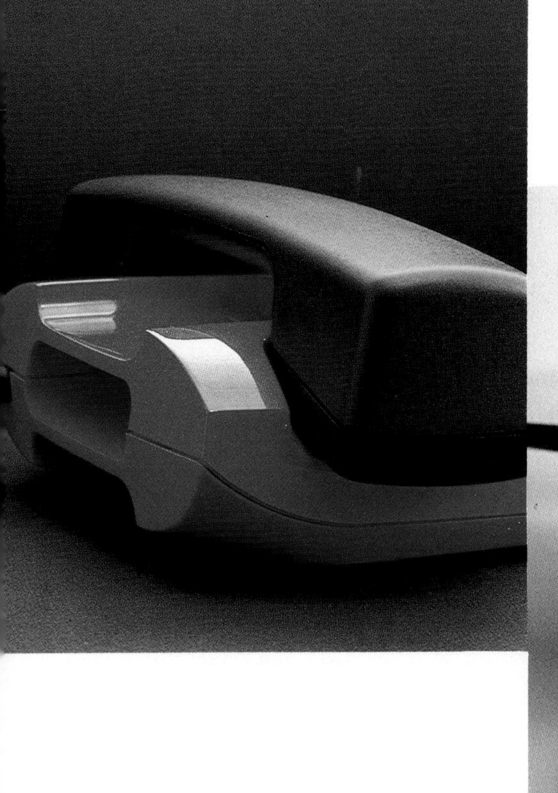

Product: Wagner Power Roller
Designers: David Miller, Robert Dawson, and Thomas Pendleton King Casey Inc. New Canaan, Connecticut
Design Firm: King Casey Inc. New Canaan, Connecticut
Client: Wagner Spray Tech Corporation Minneapolis, Minnesota
Materials: Natural and glass-filled polypropylene, glass-filled nylon, glass-filled valox

Product: Handheld Portable Electric Drill
Designer: Rudolf M. Wieland Stuttgart, West Germany
Awards: 1983 Braun Prize for Technical Design

Product:	Woodworking Safety Kit
Designers:	Chuck Haas, John Schnell, Bill Ash, David Morris, John Clark Shopsmith Inc. Dayton, Ohio
Consultant Design Firm:	Corporate Design Center Columbus, Ohio
Client:	Shopsmith, Inc. Dayton, Ohio
Awards:	1983 *Industrial Design* magazine Design Review selection
Materials:	Injection molded structural foam, natural gray finish

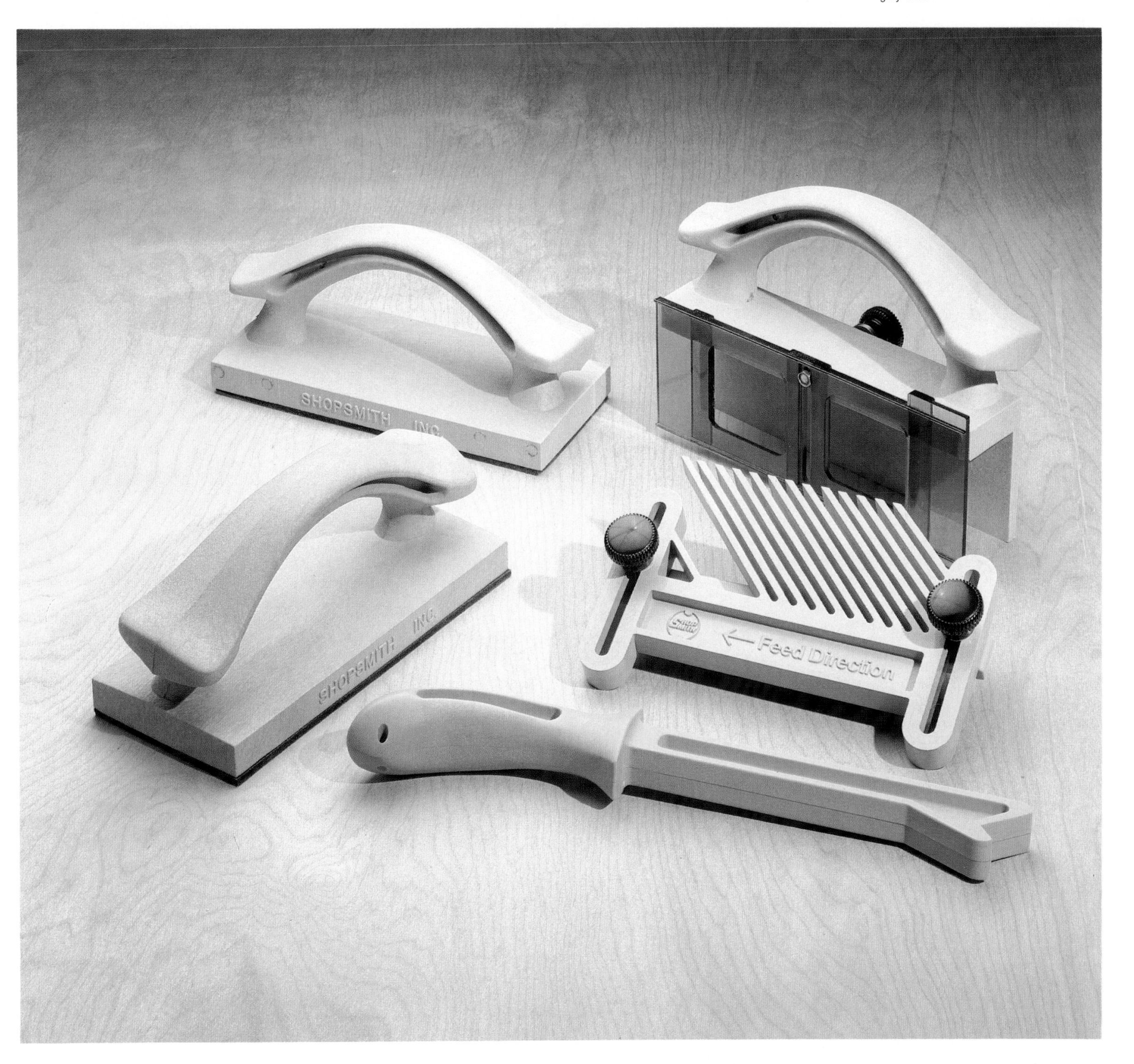

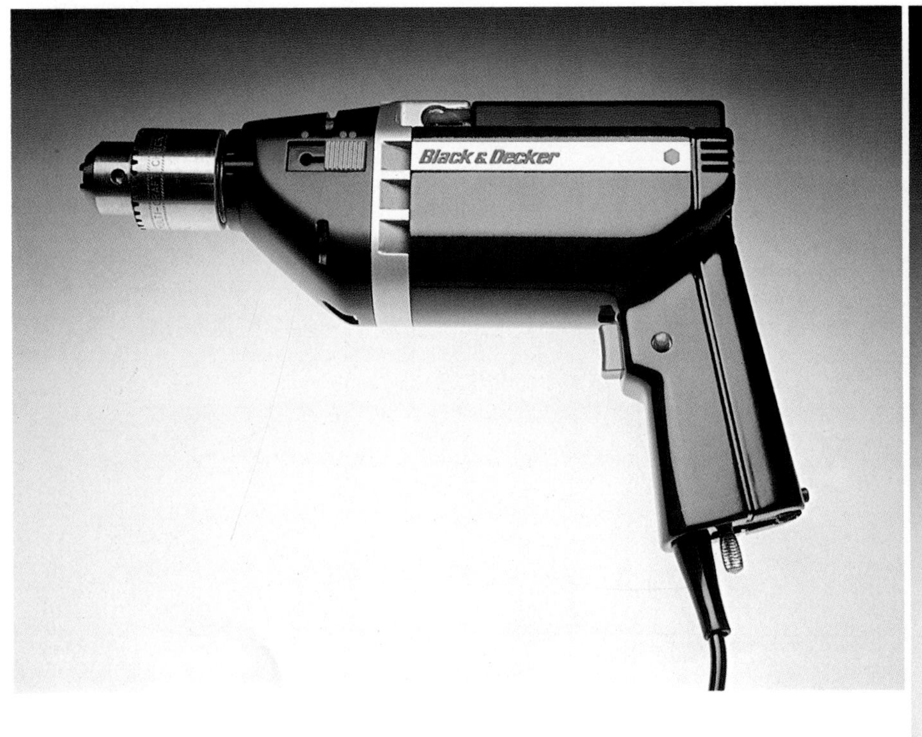

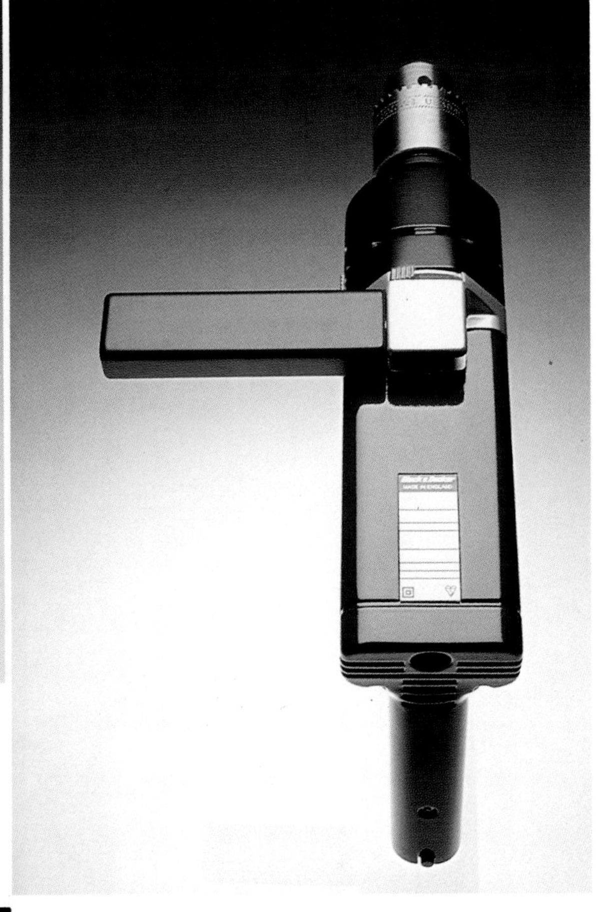

Product: Electric Drill
Design Firm: BIB Design Consultants
London, England
Client: Black and Decker Ltd.
Maidenhead, Berkshire, England

Product: Pocket Socket
Designers: Edward Levy and Michael Ballone
Fort Lee, New Jersey
Design Firm: Innovations & Development, Inc.
Fort Lee, New Jersey
Client: New Britain Tool Co., Div. of Litton Industries
New Britain, Connecticut
Materials: Injection-molded ABS handle insert molded around die-cast chrome plated steel core; carrier for six sockets is injection-molded polypropylene.

Product:	Toro Compact 50 hose/reel system
Design Firm:	King Casey Inc. New Canaan, Connecticut
Client:	Toro Company Bloomington, Minnesota

Awards:	1980 *Industrial Design* magazine Design Review selection 1980 IDSA Industrial Design Excellence Award Selected for permanent collection, Museum of Modern Art, New York, New York
Materials:	ABS plastic; Hose: polyester fabric with bonded urethane loner; brass fittings

CHAPTER 2

NCESHOUSEWARESTOOLSAPPLIANCESHOUSEWA
NCESHOUSEWARESTOOLSAPPLIANCESHOUSEWA

ECTRONICSENTERTAINMENTHOMEELECTRONIC
AINMENTHOMEELECTRONICSENTERTAINMENTH

Home Electronics and Entertainment

NGLIGHTINGLIGHTINGLIGHTINGLIGHTINGLIGHTINGLIGHTINGLIGHTINGLIGHTINGLIGHTINGLIGHTING
NGLIGHTINGLIGHTINGLIGHTINGLIGHTING

CTRESIDENTIALFURNISHINGSCONTRACTR
NTIALFURNISHINGSCONTRACTRESIDENTI

SSEQUIPMENTBUSINESSEQUIPMENTBUSINESSEQUIPMENTBUSINESSEQUIPMENTBUSINESSEQUIPMENTBUSINESSEQU
SSEQUIPMENTBUSINESSEQUIPMENTBUSINESSEQUIPMENTBUSINESSEQUIPMENTBUSINESSEQUIPMENTBUSINESSEQU

LEQUIPMENTMEDICALEQUIPMENTMEDICALEQUIPMENTMEDICALEQUIPMENTMEDICALEQUIPMENTMEDICALEQUIPMENT
ENTMEDICALEQUIPMENTMEDICALEQUIPMENTMEDICALEQUIPMENTMEDICALEQUIPMENTMEDICALEQUIPMENTMEDICAL

RIALEQUIPMENTTRANSPORTATIONINDUSTRIALEQUIPMENTTRANSPORTATIONINDUSTRIALEQUIPMENTTRANSPORTAT
RIALEQUIPMENTTRANSPORTATIONINDUSTRIALEQUIPMENTTRANSPORTATIONINDUSTRIALEQUIPMENTTRANSPORTAT

TIONALSPORTSEQUIPMENTRECREATIONALSPORTSEQUIPMENTRECREATIONALSPORTSEQUIPMENTRECREATIONALSPO
ENTRECREATIONALSPORTSEQUIPMENTRECREATIONALSPORTSEQUIPMENTRECREATIONALSPORTSEQUIPMENTRECREA

ESTEXTILESTEXTILESTEXTILESTEXTILESTEXTILESTEXTILESTEXTILESTEXTILESTEXTILESTEXTILESTEXTILES
ESTEXTILESTEXTILESTEXTILESTEXTILESTEXTILESTEXTILESTEXTILESTEXTILESTEXTILESTEXTILESTEXTILES

SFORTHEHANDICAPPEDDESIGNSFORTHEHANDICAPPEDDESIGNSFORTHEHANDICAPPEDDESIGNSFORTHEHANDICAPPED
HANDICAPPEDDESIGNSFORTHEHANDICAPPEDDESIGNSFORTHEHANDICAPPEDDESIGNSFORTHEHANDICAPPEDDESIGNS

The advent of electronics in home entertainment products has brought with it the attention of industrial designers. This does not come as any surprise—not only has the technology of electronics changed the size and shape of familiar products, but it has also generated the development of altogether new products. Both areas, clearly, are open invitations to the industrial designer.

To begin with, one of the most apparent features in contemporary consumer electronics is that they are smaller, lighter, more portable—and all the while, more durable. In terms of entertainment, at least, you *can* take it with you. And while their size and weight are diminished, so too is their price. Witness the range of new Sony products: the Walkman, Sport Walkman, and Music Shuttle all attest to the growing appeal of diminishing sizes. Likewise, the Watchman portable TV is 20 percent smaller and 20 percent lighter than its predecessor as a result of flat display picture tube technology.

Advanced electronics also give the consumer a wide range of product choices. Like most contemporary business products, the sophistication of home entertainment products is usually determined by the degree of control they offer the user. The Technics SL-QL15 turntable, for example, invites the user to "step into the world of programmable convenience" by selecting both the songs and the order in which they are to be played. An LED confirms the selection, and a repeat key will repeat the series. Songs within the program can also be skipped, repeated, or canceled in "random access programming" which is achieved by microcomputer technology. While this degree of programmable convenience may be necessary or convenient only to the very few, it is nevertheless what distinguishes "state of the art" consumer electronics.

Similarly, the ADS Atelier series of audio components offers a receiver which provides for connecting two tape decks and two sets of speakers, and allows the user to copy tapes while listening to the radio or playing a record. But the ADS series also demonstrates how the role of the designer is changing. Because electronics is itself such a recent product category, its design and manufacturing procedures have not yet become standardized. There is enough flexibility for the designer to have an integral role in actual product development, as indeed, in this case, Dieter Rams did. Collaborating with engineering specialists, Rams and the Braun electronics team were "responsible for helping conceive the various capabilities built into the products as well as extending these in a logical manner to the exterior envelope to ultimately contribute to accessibility and ease of use." That designers have recognized electronics as uncharted territory which invites and will accommodate their participation at much earlier stages benefits both their profession and engineering. More to the point, perhaps, is that the collaboration between the two signals a broader, more research-oriented approach to product development.

While this designer participation may result, in general, in a more integrated product, it may lead more particularly to a product that is human-engineered. The industrial designer is more likely than the engineer to consider how the user and the product will come to terms with one another. In Digital's Professional 350 personal computer, for example, the "human factors" research of the designers helped to specify at an early stage the angle of the tilt for the CRT, the angle for keyboard controls, mass storage access points, installation studies, cooling, internal component configurations, and cable management—as well as color, form, texture, material, manufacturing processes, and overall systems image—all the more traditional domain of the industrial designer. The system that evolved from this collaboration is indeed "an expandable system approach," which can accommodate a variety of personal computing capabilities, for telephone and database management to accessing to larger computing systems for executives and small businesses.

While the term "user-friendly" may have become somewhat stale, electronics engineers who continue to recognize the value of human factors input in the design of their equipment will continue to enjoy the edge. Despite their high visibility, electronics products remain unfamiliar—and, often, threatening enough to the general consumer that the convenient placement of their controls, absence of intimidating hardware, clear and informative graphics, and overall accessible appearance remain vital to their adoption and use. And as the capabilities of such equipment grow and become more complex, this will become all the more true.

By inviting designer participation at an earlier stage, particularly in new products, the electronics industry also introduces a whole new set of problems. The design of the compact disc player is a case in point. With the advent of digital audio recording, fidelity has reached a near perfect degree of accuracy. Digital audio is an outgrowth of computer technology which records, stores, and reproduces sounds by first giving them numerical codes. These codes are assigned to each "sample" of music being recorded, with samples taken 44,100 times each second. Digital signals engraved on a compact disc in the form of microscopic pits are then read by a laser beam (eliminating, of course, the need for the conventional needle, or stylus). The disc itself is only a little less than five inches in diameter.

What all this adds up to for the designer is that the new technology, aside from promoting miniaturization, also removes the traditional constraints of shape. Not only can the product be much smaller, but it doesn't necessarily need to be housed in the rectangular box of the traditional turntable and amplifier. Although this is certainly liberating to some degree, designers who "liberate" their products entirely from these conventions find themselves at a disadvantage. That is: the consumer looks for the familiar and searches out the recognizable. Even if the technology is brand new, if it *looks* brand new it is more than likely to put the consumer off. Products that look as though they do what they do—in the eyes of the consumer—project a sense of logic which is reassuring. And new technology, when presented as such, can be intimidating. Thus, the design of digital audio equipment which takes conventional housing of audio equipment into account may meet with a better response than units which *look* as new as they really are. While ironies are implicit here, designers who recognize them—and the delicate balancing act they necessitate— perhaps achieve best results.

In concluding, it may be helpful to point out that some of the products shown here could have been shown as easily, or as appropriately, in other categories. The personal computers, most notably, are intended as much for use in small businesses as in the home. That several *are* included here only reiterates a point made elsewhere in this book—that as more and more people choose to work in their homes, equipment and furnishings for home and for office become less distinguishable. This is also why Richardson/Smith's 1982 computer prototype designed for *Time* magazine's "Machine [rather than Man] of the Year" is included here. As the computer becomes standard equipment for home as well as office, its design becomes more humanized. Hence the tilting and swiveling monitor, the adjustable, thin keyboard, and lively graphics that often serve more than one function. Designers for home electronics and entertainment products who note that the fine line between home and office is apt to become even finer are clearly recognizing their market.

If any overall conclusion is to be drawn from all this, it is that product design of this category demonstrates a more integrated, collaborative approach among designers, engineers, and technicians than in the past. Simply on account of its recent emergence, electronics and the development of electronic products have made room for the designer. Designers who have recognized this and responded by contributing to early product development are not only designing products that are as easy to operate as they are to look at; they are also setting new precedents for their colleagues working in other areas of design.

Product: JBL Auto Speakers
Designers: David Muramatsu, Tom Rupp, Luis Urquidi
Orange, California
Design Firm: Halsted & Muramatsu, Orange, California
Client: JBL Sound, Northridge, California
Awards: 1983 *Industrial Design* magazine Design Review selection
Materials: Frame: brushed aluminum with black-epoxy metal powder paint coating; Logo: die-cast with JBL orange paint

Product: MCS Stereo Receiver 3285
Designers: Jan Tribbey and Rick Blanchard, JC Penney
New York, New York
Client: JC Penney, New York, New York
Awards: 1983 *Industrial Design* magazine Design Review selection
Materials: Cabinet: vinyl-wrapped steel; facade: injection-molded ABS; control panel and switch caps: natural finish aluminum

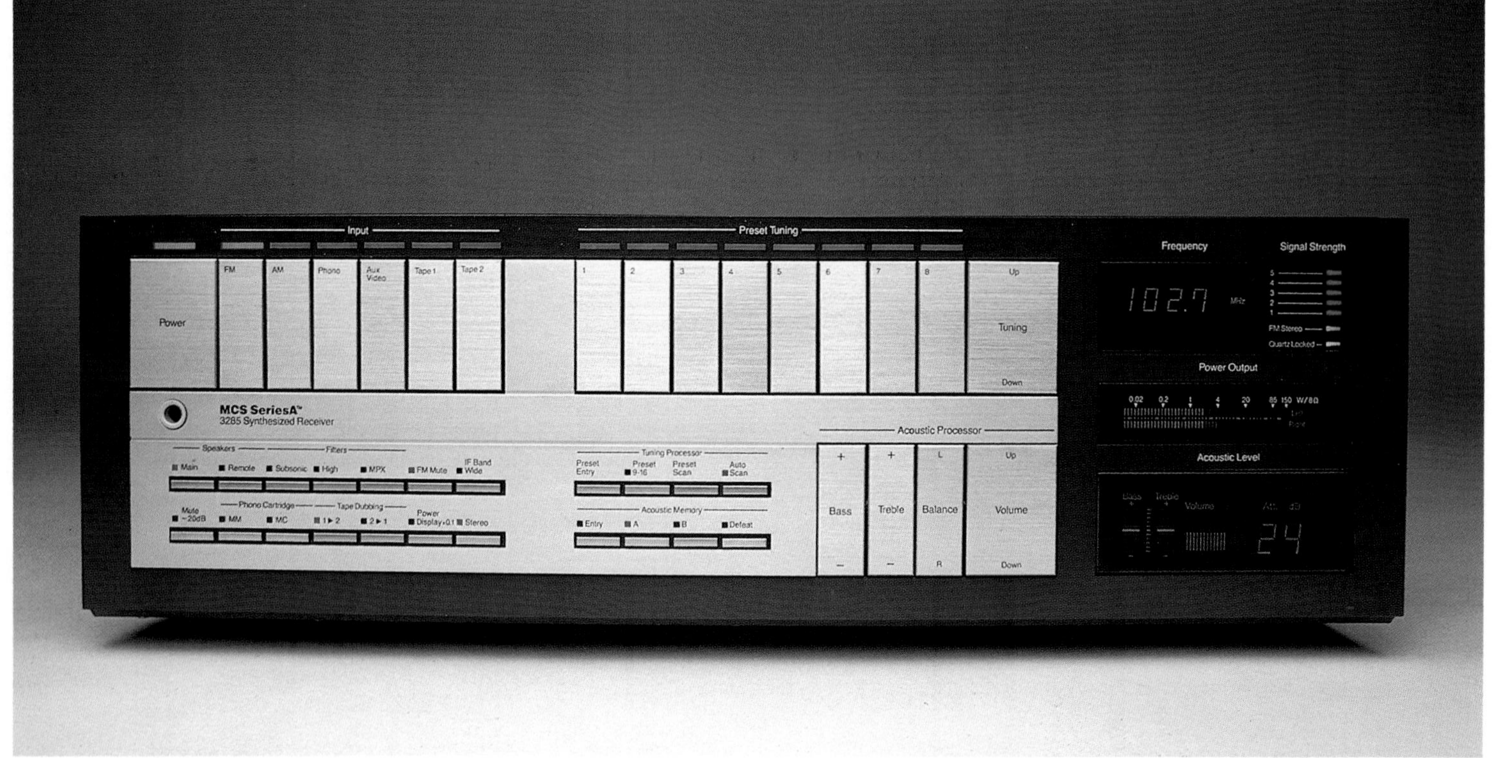

Product:	SL-QL15 Programmable Quartz Linear Tracking Turntable
Designer:	Engineering staff, Matsushita Electrical and Industrial Company of Japan, Osaka, Japan
Client:	Technics; Panasonic Company Division of Matsushita Electric Corporation of America Secaucus, New Jersey

Product:	MCS Turntable 6730
Designers:	Jan Tribbey and Rick Blanchard, JC Penney New York, New York
Client:	JC Penney, New York, New York
Awards:	1983 *Industrial Design* magazine Design Review selection
Materials:	Injection molded acrylic lens and dust cover; Feet: spring wrapped rubber; Base: injection molded acrylic and low-resonance foam; Control pads: injection molded ABS

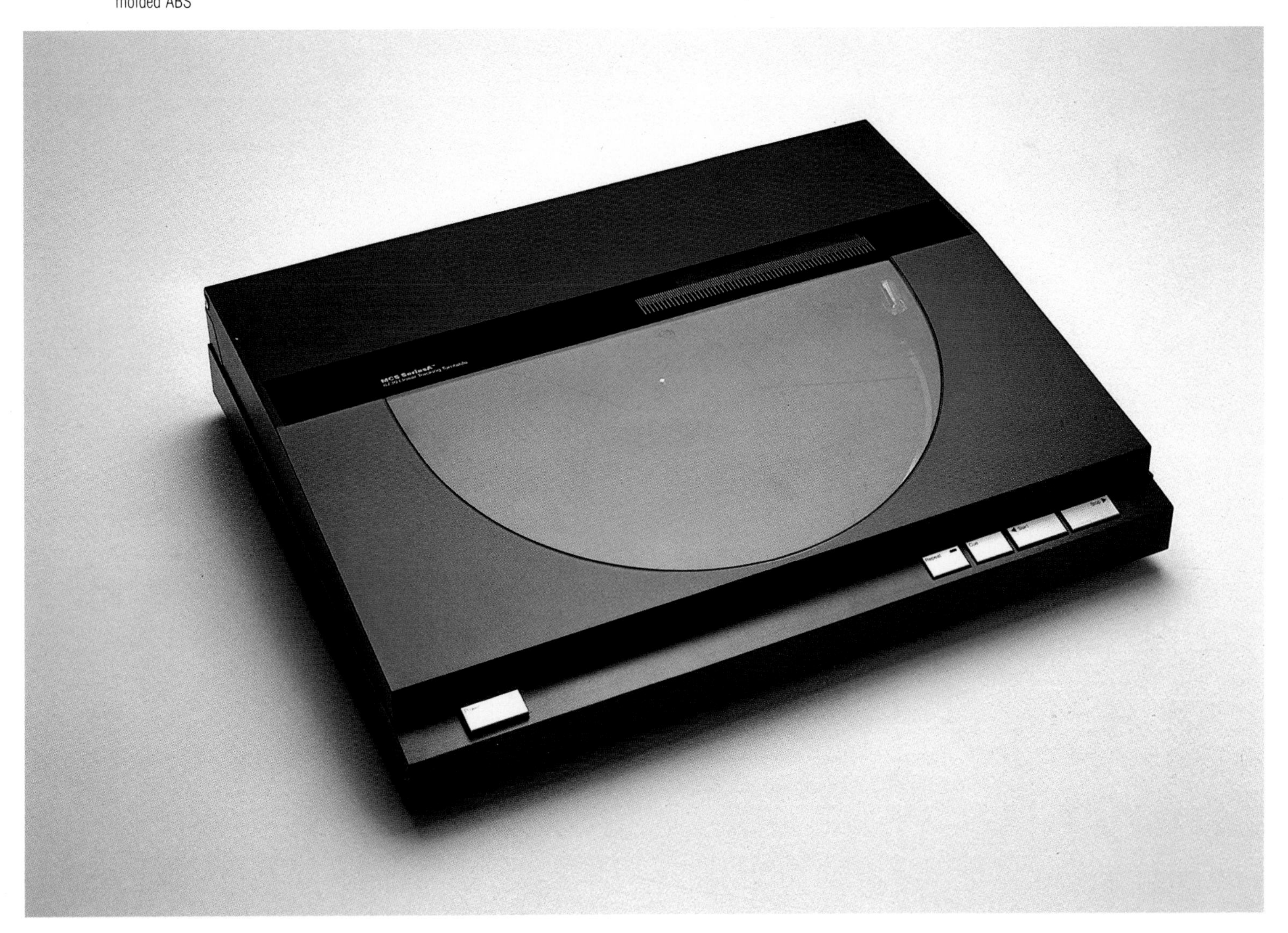

Product: Concept 51 K
Design Firm: Frogdesign, Campbell, California
Client: Wega Electronic, West Germany
Materials: Polycarbonate housing

Product: MCS 1210 Stereo System
Designers: Jan Tribbey and Rick Blanchard, JC Penney;
Chris Hacker and Martin Thaler, Lee Manners and Associates, New York, New York
Client: JC Penney, New York, New York
Awards: 1983 *Industrial Design* magazine Design Review selection
Materials: Cabinets: particleboard with silver metallic vinyl wrap; Cabinet fronts: silver metallic injection-molded ABS; Speaker grilles: vacuum-formed ABS

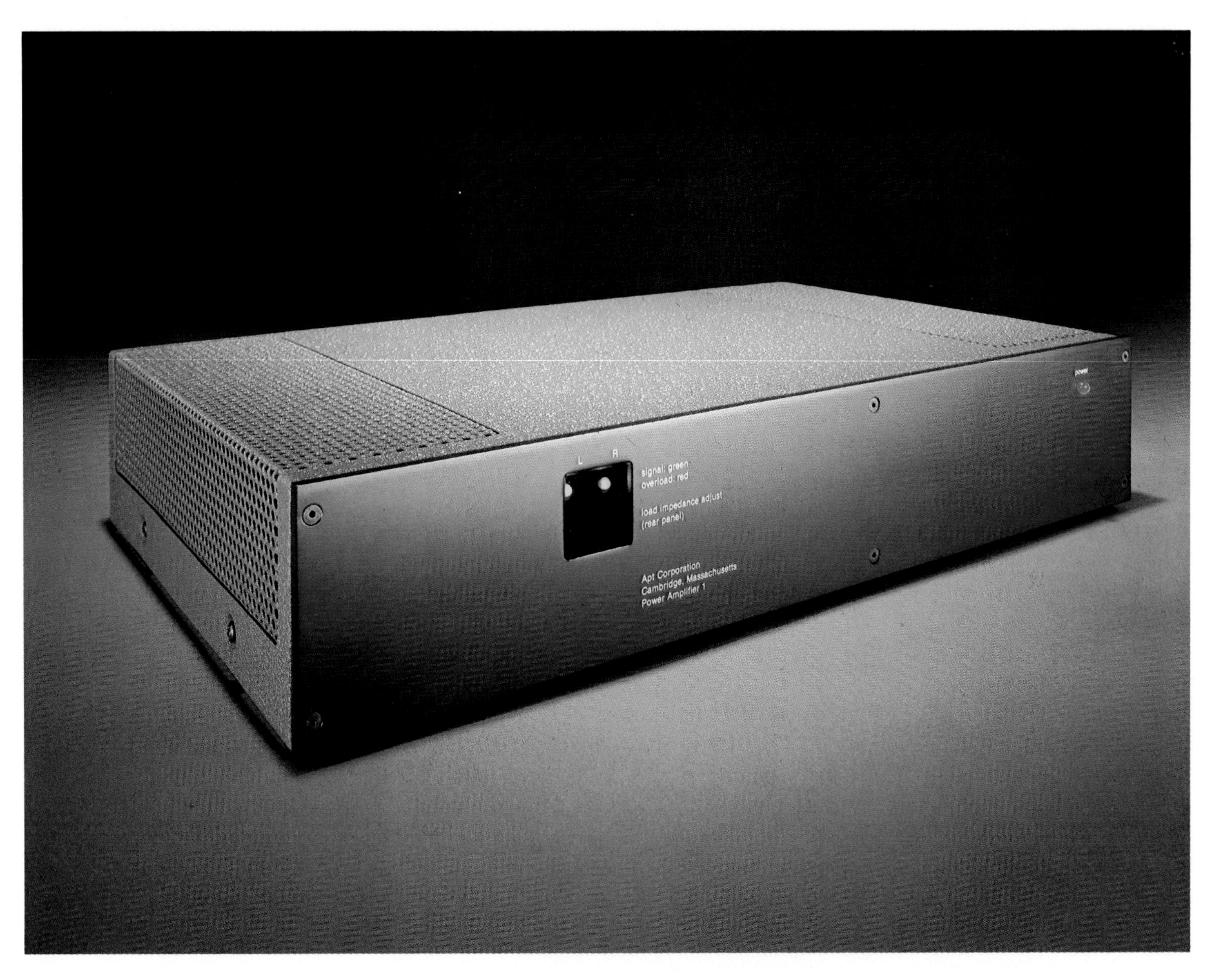

Product: Apt 1 Stereo Power Amplifier
Designer: Charles Rozier, New York, New York
Client: Apt Corporation, Cambridge, Massachusetts
Awards: 1981 *Industrial Design* magazine Design Review selection
Materials: Formed steel case with gray wrinkle finish enamel; painted aluminum front

Product: Kyocera DA-01 Compact Disc Player
Designer: Engineering division, Kyocera International Inc.
Client: Kyocera International Inc., Warren, New Jersey
Materials: Steel frame; Front panel: aluminum with gunmetal finish; exterior wood panels

Product: Beocenter 7700
Designer: Design department, Bang & Olufsen A/S
Struer, Denmark
Client: Bang & Olufsen A/S, Struer, Denmark
Materials: Injection molded chassis

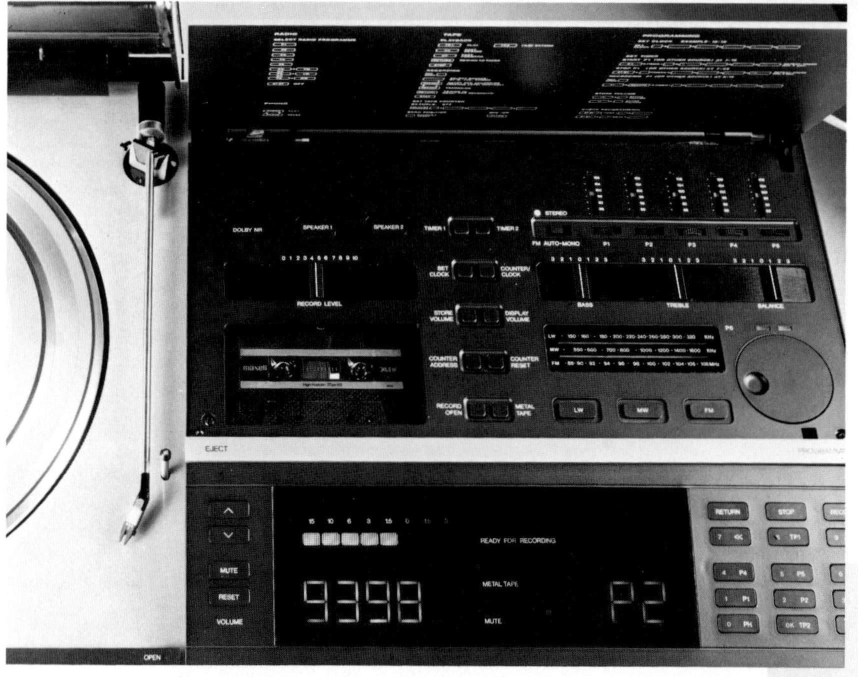

Product: XRM-10 Music Shuttle
Designer: Sony Tokyo Design Engineers
Client: Sony Corporation, Tokyo, Japan

Product: ADS Atelier Audio Components
Designer: Dieter Rams
Clients: ADS, Analog & Digital Systems, Inc., Wilmington, Massachusetts, and Braun Electronic GmbH Kronberg, West Germany
Materials: Bevelled matte black components can be stacked or set side by side

Product: FD-20A Watchman
Designer: Sony Tokyo Design Engineers
Client: Sony Corporation, Tokyo, Japan

Product: FD-30A Watchman
Designer: Sony Tokyo Design Engineers
Client: Sony Corporation, Tokyo, Japan

Product: Sony Sports Walkman
Designer: Sony Tokyo Design Engineers
Client: Sony Corporation, Tokyo, Japan

Product: Steinberger Bass
Designer: Ned Steinberger, Newburgh, New York
Client: Steinberger Sound Corp., Staten Island, New York
Awards: 1981 IDSA Industrial Design Excellence Award
1981 *Time* Magazine selection for five best of the year
Materials: Neck and body: graphite-fiber and glass-fiber reinforced epoxy resin and polyester gel coat finish; solid brass and stainless steel hardware

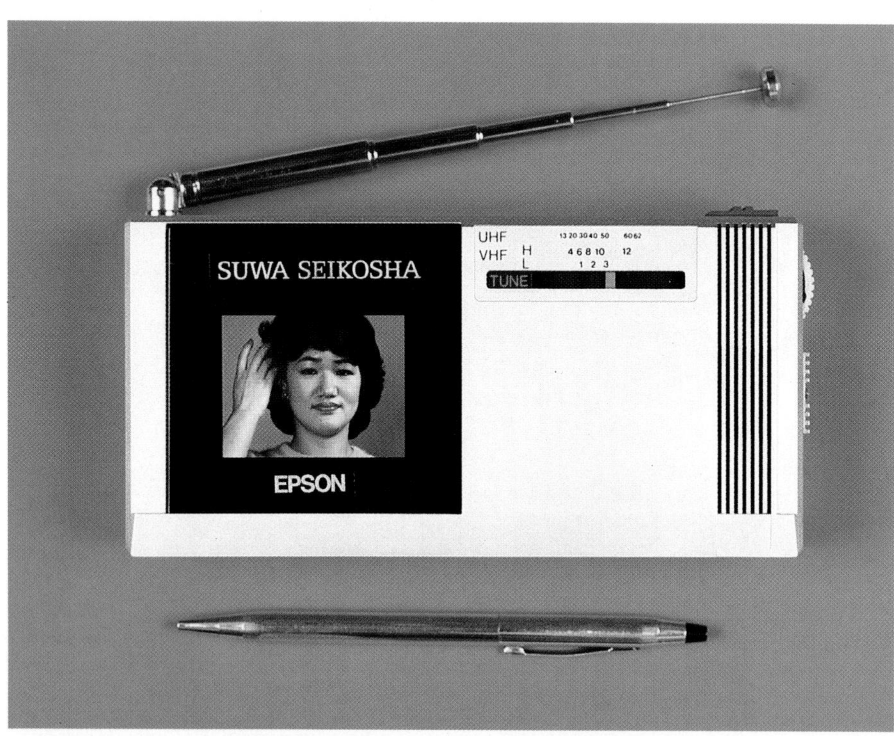

Product: Seiko Pocket Color TV
Manufacturer: Seiko, Suwa Seikosha Co., Ltd., Suwa, Japan

Product: Microtek 36cm remote control portable television
Designers: H. Veaudry, M. A. Carr, C. Cunningham, and K. Chadwick and staff designers at Tek Electronics Ltd.
Client: Tek Electronics (pty) Limited Wilsonia, South Africa
Awards: 1982 Shell Design Award, consumer products
Materials: ABS housing

Product:	FROGLINE Television Modules
Design Firm:	frogdesign, Campbell, California
Client:	Sony Corporation, Tokyo, Japan
Awards:	Excellence of Design Award, Stuttgart Design Center
Materials:	Polycarbonate and polystyrene

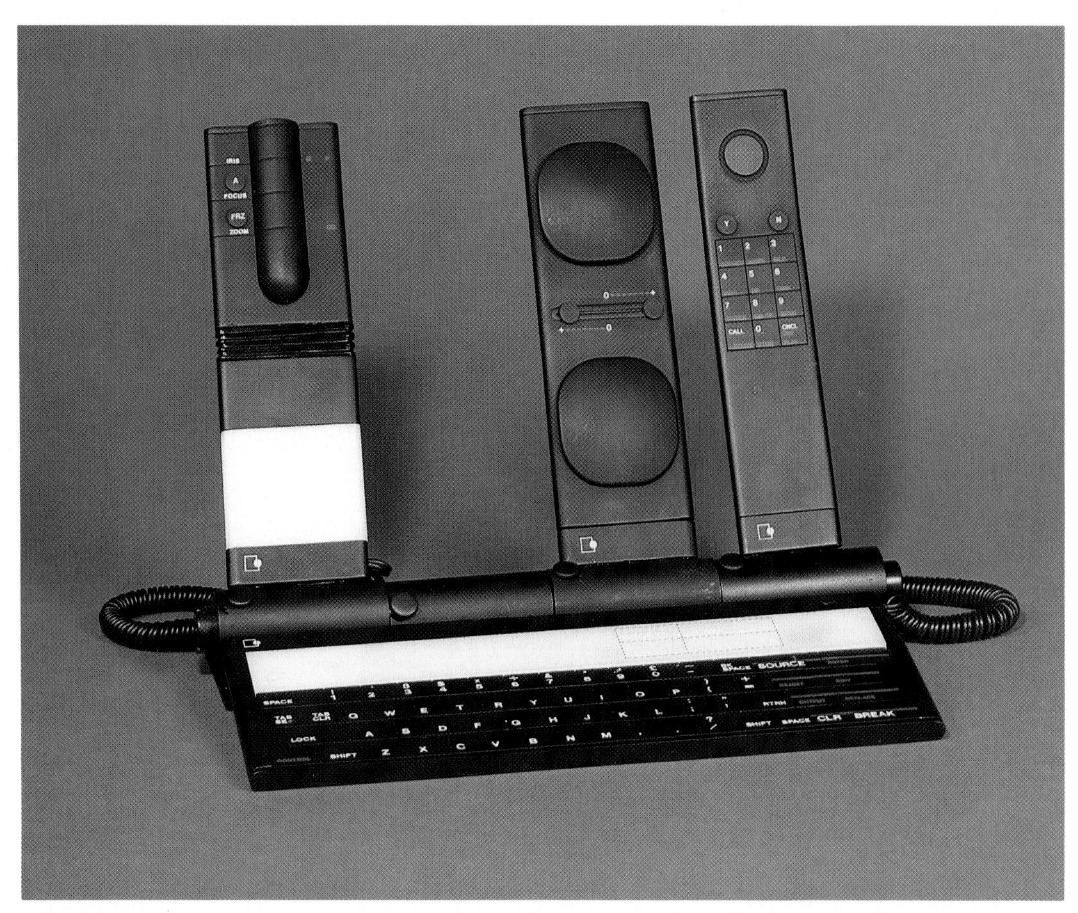

Product:	Home computer prototype
Designer:	Gustavo Rodriguez, Chicago, Illinois
Awards:	1983 Braun Prize for Technical Design

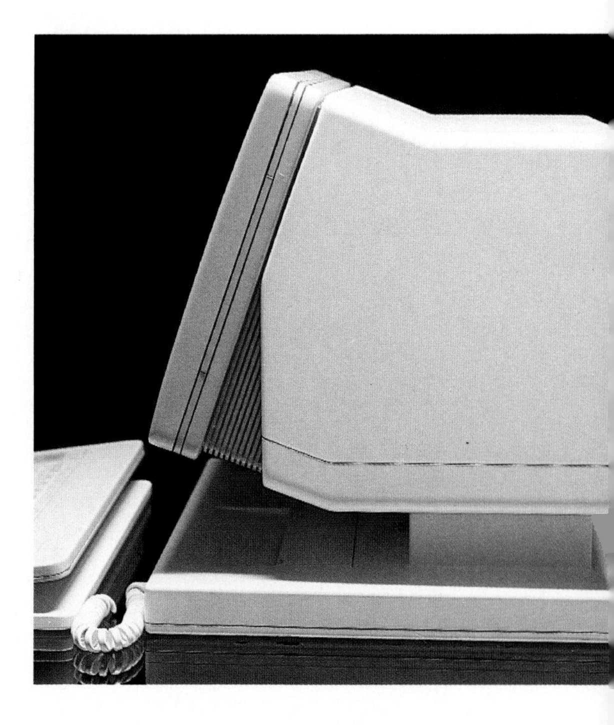

Product:	IBM PCjr
Designer:	IBM Entry Systems Division, Boca Raton, Florida
Client:	IBM Corporation, Armonk, New York
Materials:	Polycarbonate housing

Product: Digital 350 Professional Series
Designer: Industrial Design Group, Digital Equipment
Maynard, Massachusetts
Client: Digital Equipment, Maynard, Massachusetts
Awards: 1983 *Industrial Design* magazine Design Review selection
Materials: All component housing injection molded in Noryl N-190; texture mold Tech 1013

Product: Personal computer prototype designed to represent "The Machine of the Year" for *Time* magazine
Deisgner: Ed Lawing, Richardson/Smith
Worthington, Ohio
Design Firm: Richardson/Smith, Worthington, Ohio
Client: *Time* magazine, *Time*, Inc.
New York, New York
Materials: Acrylic and wood mockup

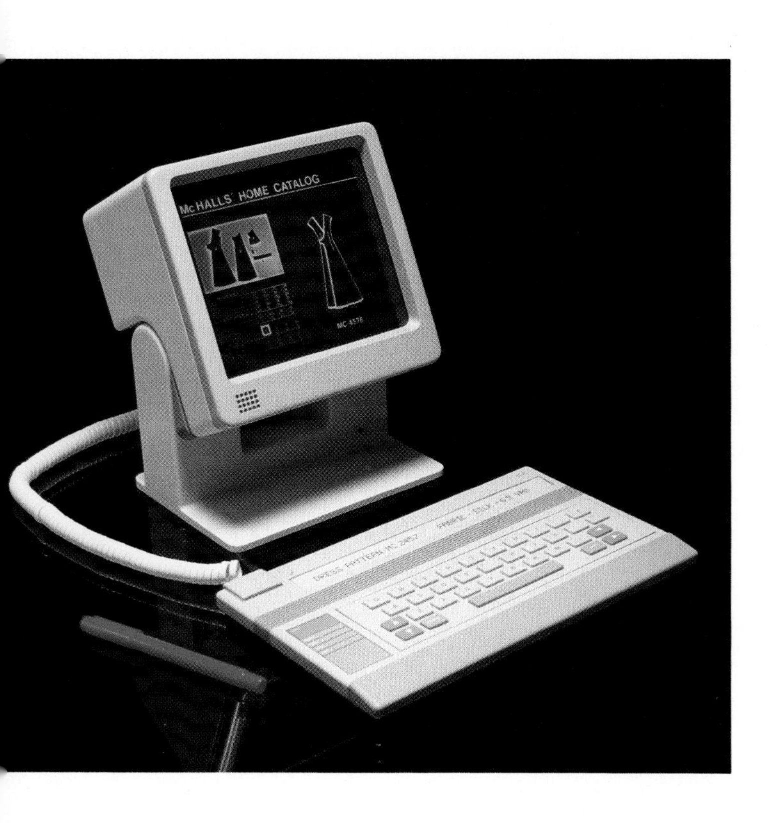

CHAPTER 3

CESHOUSEWARESTOOLSAPPLIANCESHOUSEWARESTOOLSAPPLIANCESHOUSEWARESTOOLSAPPLIANCESHOUSEWAREST
CESHOUSEWARESTOOLSAPPLIANCESHOUSEWARESTOOLSAPPLIANCESHOUSEWARESTO

Lighting

CTRONICSENTERTAINMENTHOMEELECTRONICSENTERTAINMENTHOMEELECTRONICSE
INMENTHOMEELECTRONICSENTERTAINMENTHOMEELECTRONICSENTERTAINMENTHOM

GLIGHTINGLIGHTINGLIGHTINGLIGHTINGLIGHTINGLIGHTINGLIGHTINGLIGHTINGLIGHTINGLIGHTINGLIGHTING
GLIGHTINGLIGHTINGLIGHTINGLIGHTINGLIGHTINGLIGHTINGLIGHTINGLIGHTINGLIGHTINGLIGHTINGLIGHTING

TRESIDENTIALFURNISHINGSCONTRACTRESIDENTIALFURNISHINGSCONTRACTRESIDENTIALFURNISHINGSCONTRA
TIALFURNISHINGSCONTRACTRESIDENTIALFURNISHINGSCONTRACTRESIDENTIALFURNISHINGSCONTRACTRESIDE

SEQUIPMENTBUSINESSEQUIPMENTBUSINESSEQUIPMENTBUSINESSEQUIPMENTBUSINESSEQUIPMENTBUSINESSEQU
SEQUIPMENTBUSINESSEQUIPMENTBUSINESSEQUIPMENTBUSINESSEQUIPMENTBUSINESSEQUIPMENTBUSINESSEQU

EQUIPMENTMEDICALEQUIPMENTMEDICALEQUIPMENTMEDICALEQUIPMENTMEDICALEQUIPMENTMEDICALEQUIPMENT
NTMEDICALEQUIPMENTMEDICALEQUIPMENTMEDICALEQUIPMENTMEDICALEQUIPMENTMEDICALEQUIPMENTMEDICAL

IALEQUIPMENTTRANSPORTATIONINDUSTRIALEQUIPMENTTRANSPORTATIONINDUSTRIALEQUIPMENTTRANSPORTAT
IALEQUIPMENTTRANSPORTATIONINDUSTRIALEQUIPMENTTRANSPORTATIONINDUSTRIALEQUIPMENTTRANSPORTAT

IONALSPORTSEQUIPMENTRECREATIONALSPORTSEQUIPMENTRECREATIONALSPORTSEQUIPMENTRECREATIONALSPO
NTRECREATIONALSPORTSEQUIPMENTRECREATIONALSPORTSEQUIPMENTRECREATIONALSPORTSEQUIPMENTRECREA

STEXTILESTEXTILESTEXTILESTEXTILESTEXTILESTEXTILESTEXTILESTEXTILESTEXTILESTEXTILESTEXTILES
STEXTILESTEXTILESTEXTILESTEXTILESTEXTILESTEXTILESTEXTILESTEXTILESTEXTILESTEXTILESTEXTILES

FORTHEHANDICAPPEDDESIGNSFORTHEHANDICAPPEDDESIGNSFORTHEHANDICAPPEDDESIGNSFORTHEHANDICAPPED
ANDICAPPEDDESIGNSFORTHEHANDICAPPEDDESIGNSFORTHEHANDICAPPEDDESIGNSFORTHEHANDICAPPEDDESIGNS

For centuries people have been devising comfortable places to sit, convenient surfaces to put things on, and tools that will refine and extend their reach and capabilities. By design, we have made our world more accommodating. Not until much more recently, however, have we been able to enhance how we actually see the world. That is, not until the relatively recent discovery of electricity and the invention of artificial light.

Within this relatively recent history, lighting design has referred largely to the aesthetics of the fixture rather than to the actual quality of illumination. It is simply the short history, perhaps, that accounts for the narrow vision and definition of lighting design. But, for two reasons, lighting designers have finally come to consider the quality of illumination integral to lighting design. First is that recent and advanced lighting technology—most notably of enhanced fluorescents and tungsten-halogen lamps—has permitted and encouraged a discrimination in actual lighting sources as well as the lamps that house them. And second is the simple fact that seven-eighths of all our perceptions come through sight. With the available technology, designers can no longer afford to overlook, so to speak, how sight can be enhanced. So the current innovations in lighting are not necessarily an additional technology, but in the applications of existing technology.

American and European lighting designers have had two very different approaches to these applications. European designers have tended to determine the lighting requirements of a space before furnishing it. Americans, on the other hand, have been more prone to decide what needs to be lit before determining how it can be lit. Assuming that the light source, distribution, and placement of the lamp are the three foremost questions, they are ones most American designers feel cannot be answered until the natural daylighting and spatial relationships within the space have been determined. A fluorescent light, for example, will change the color of whatever it falls on, hence it would be foolish to decide to use a fluorescent lamp without knowing what, exactly, it was meant to light.

Nevertheless, this approach has all too often designated lighting as a secondary design element. It is from this deferential position that lighting designers have only recently begun to rescue their material. Architects and designers are beginning to accept notions well known to designers in the theater: Lighting affects mood and behavior; it can generate emotional and psychological responses ranging from apprehension and terror to tranquility and contentment. And although the office and home may be the backdrop for enactments usually less dramatic than those occurring on the stage, how these spaces are lit is surely as vital to the mood they finally project. Just as the socket and single bulb solution to lighting gave way to track lighting in the 1960s, the all-in-one, everything-at-once track fixture perhaps sacrificed the quality of illumination for convenience. It is now being replaced by more thoughtful applications that consider different lighting sources and purposes.

In determining the light source, the bearing of light upon color is perhaps most important. Incandescent lightbulbs produce a glow similar to that of afternoon sunlight: The light is the result of an object being heated to the degree at which it glows. Whether it comes from the sun or an electric bulb, the light has a continuous spectrum that begins at a warm red and shifts to a blue white as the heat intensifies. It is the richness in reds, however, that give the bulb its warm red glow which appeals to most people because it enhances skin tone. Early fluorescent lights were deficient in reds, peaking instead in yellows and greens. Fluorescent lights now, however, have a much broader spectrum, more complete in blue and red areas and much closer to natural daylight. The light from some fluorescents now can resemble the warm reds of incandescents or late afternoon sunlight while the light emitted from others is closer to cooler, white midday light. Given the energy efficiency and practicality of fluorescents—long bulb life and diminished heat output—their increased application will perhaps have less to do with the quality of illumination they are able to achieve and more to do with putting aside traditional notions that fluorescent light is unnatural and thus unhealthy.

As much a variable as color is distribution: It can be direct, with 90 to 100 percent of its output downward; semidirect, with 40 to 60 percent of its output downward; diffused, with upward and downward output nearly equal; or indirect, with most of the light directed upward.

As designers become aware of the fact that lighting applications for home and office can have control and accuracy, they recognize the range of sources available to them and, indeed, lighting becomes a design element in its own right. The differences between incandescent and fluorescent lighting can highlight and dramatize one another: Because fluorescent light comes from a gas, it can appear to move and can be used in a dynamic composition with the more static light emitted from an incandescent. Or, artificial light can be selected to correspond to the natural daylighting of a space in a flexible—and energy efficient—arrangement that recognizes both time of day and time of year.

Tungsten-halogen lamps, often known as quartz, offer a new realm of possibilities. Although they are incandescent, they appear to produce more light from a smaller source. (Actually, it is just more concentrated.) They also produce a slightly whiter light than the standard incandescent, and the bulb has a longer life.

If current lighting innovations are to be found in applications rather than new technology, no place presents a more imposing challenge than the electronic office. Especially with the ever present CRT screen, quantity of light becomes less of an issue than quality. The greatest problem is the standard recessed lighting found in most offices that exacerbates screen glare and creates distracting shadows on the keyboard. As lighting consultant James Nuckolls points out, "The distribution of specially designed assymetric task lights offered by several manufacturers were developed for flat paperwork; they added more glare than they prevented on the multipled angled surfaces of the CRT installations." Nuckolls continued to discover in preliminary studies that varied intensities were preferred to general illumination at one level. That is, copy stand, screen surface, background, keyboard area, and flat work area were best illuminated separately. He also found that cool tints of white light were preferred, that the light should come from above and slightly in front of the worker, and that the light on the background area was most effective when distributed unevenly. While these findings were preliminary, they do address the problems of eye strain and fatigue suffered by many CRT users. And, they reestablish the fact that the lighting requirements of the electronic workstation can be met by new applications rather than new technology.

Both the energy consciousness of the 1970s and the electronics of the 1980s have forced manufacturers and consumers to become innovative in searching out these new applications. As they continue to find them, lighting is apt to become a primary, rather than secondary, design element. Most conspicuous about contemporary lighting, perhaps, is that it is in this period of transition.

Product:	BEGA Downlight
Design Firm:	BEGA Menden, West Germany
Awards:	iF Exhibition, "Good Industrial Design," Hanover 1983
Materials:	Aluminum alloy and stainless steel

Product: Wall Lamp
Designers: King, Miranda, Arnaldi
Milan, Italy
Client: Arteluce
Milan, Italy
Awards: 1982 IBD Product Design Gold Award
Materials: Cast glass diffusor, metal wall bracket

Product: Aurora Borealis
Designer: Koen de Winter
Beaconsfield, Quebec, Canada
Client: Danesco, Inc.
Montreal, Quebec, Canada
Awards: 1982 Design Canada Award
Materials: Aluminum, baked epoxy

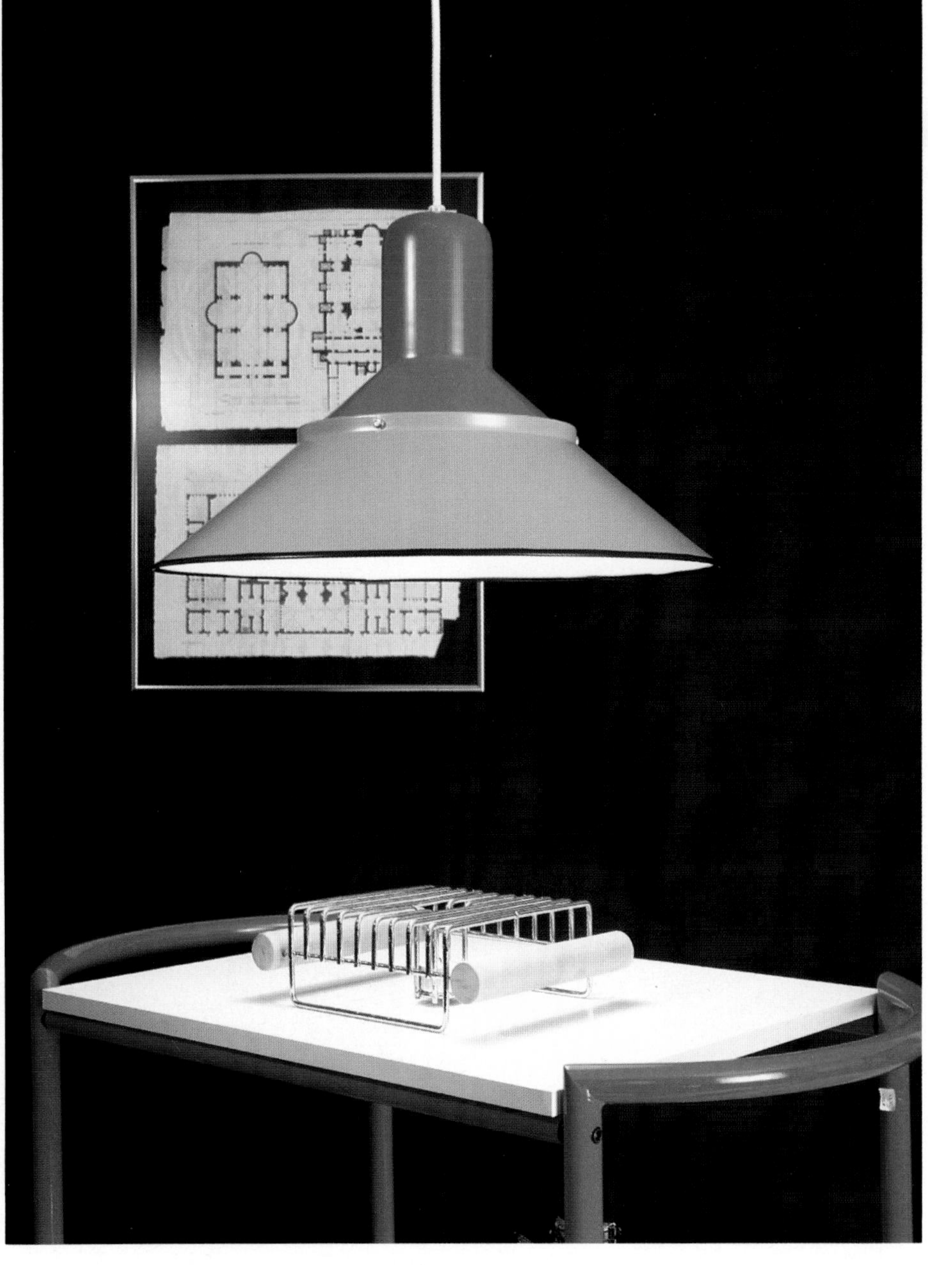

Product:	Aurora Borealis
Designer:	Koen de Winter Beaconsfield, Quebec, Canada
Client:	Danesco, Inc. Montreal, Quebec, Canada
Awards:	1982 Design Canada Award
Materials:	Aluminum, baked epoxy

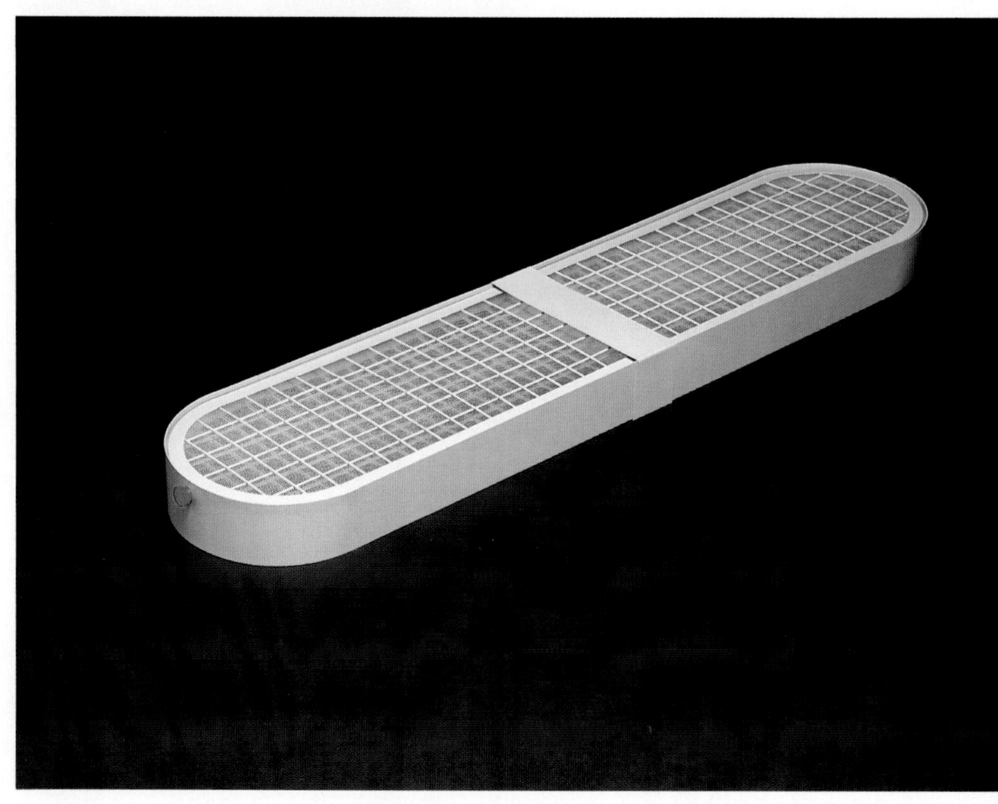

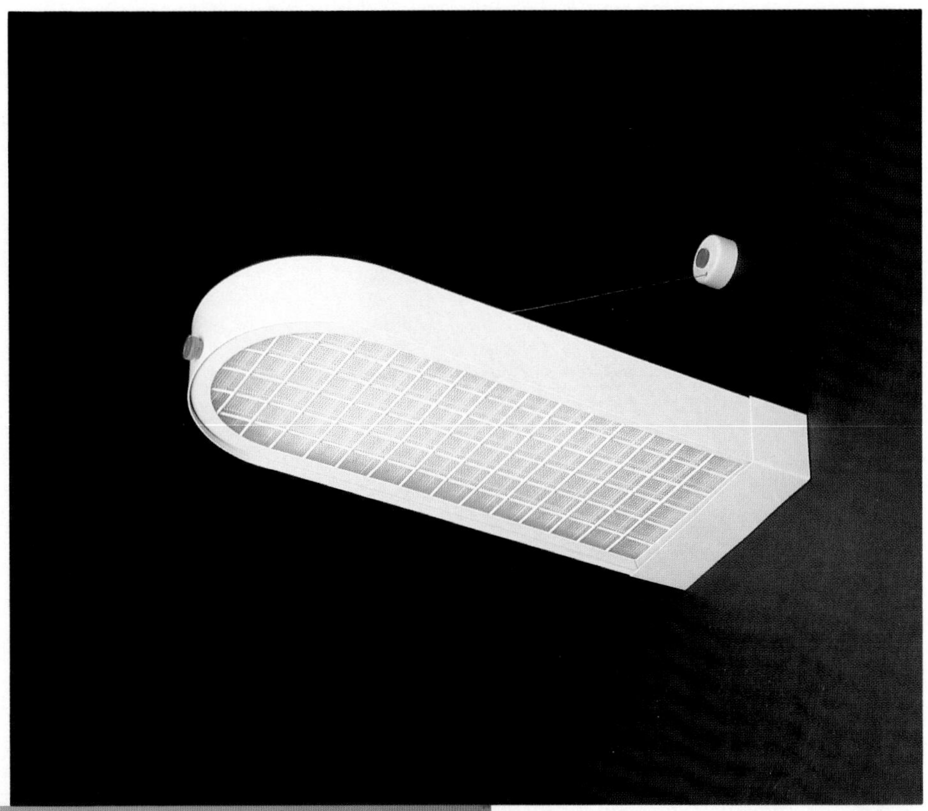

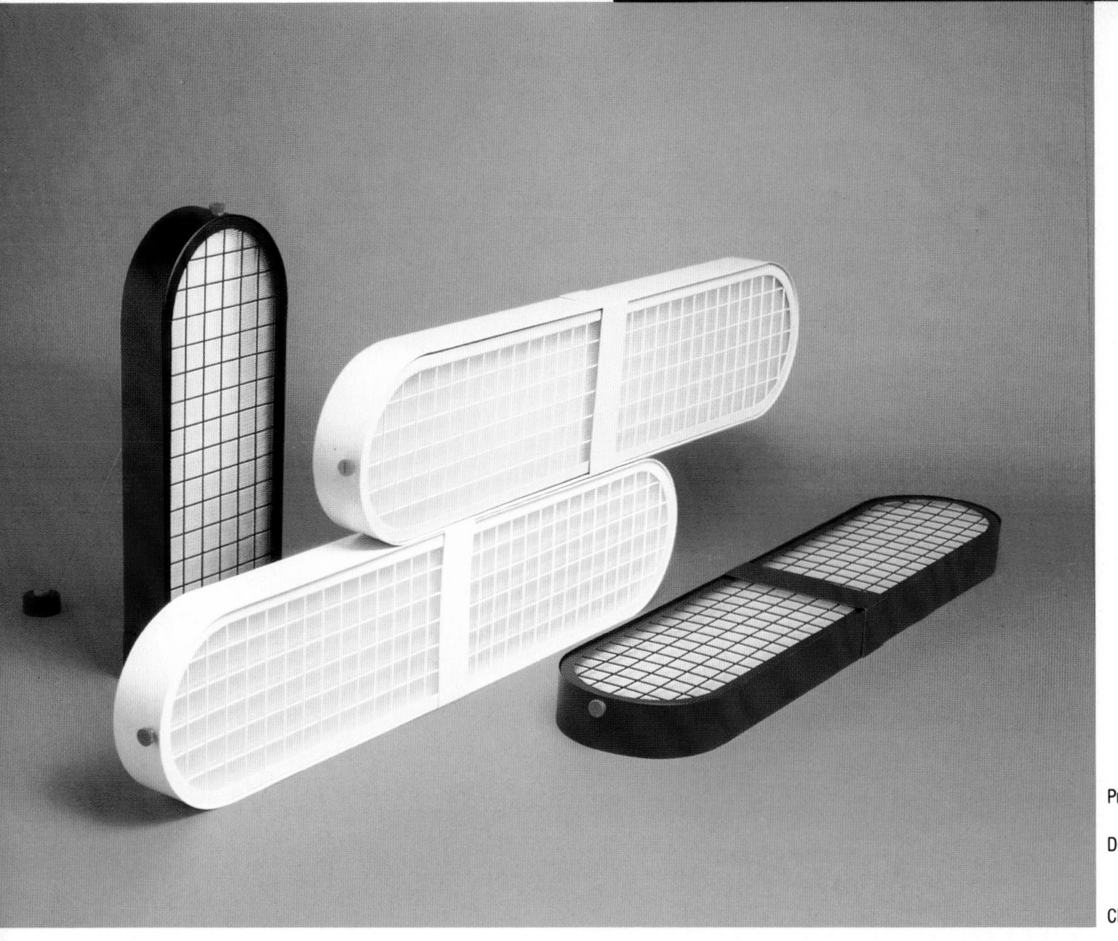

Product:	Mary (single tube) and Mary Sue (double tube)
Design Firms:	Walker/Group, Inc. and David A. Mintz, Inc. New York, New York
Client:	Lightron-of-Cornwall, Inc. New Windsor, New York
Awards:	1983 IBD Product Design Gold Award
Materials:	Cast aluminum steel wire grid, plastic refractor, fluorescent bulbs

Product:	Lamps
Designer:	Alvar Aalto (ca. 1940s and 1950s)
Reproduced by:	International Contract Furnishings New York, New York
Materials:	Painted brass and iron

Product:	Aurora Borealis
Designer:	Koen de Winter Beaconsfield, Quebec, Canada
Client:	Danesco, Inc. Montreal, Quebec, Canada
Awards:	1982 Design Canada Award
Materials:	Aluminum, baked epoxy

Product: Lamp (ca. 1952)
Designer: Alvar Aalto
Reproduced by: International Contract Furnishings
New York, New York
Materials: Brass; inside painted white

Product: Cyclos
Designer: Michele de Lucchi
Milan, Italy
Client: Artemide Inc.
New York, New York
Materials: Gray painted metal structure with partial frosted glass diffusor, fluorescent bulb

Product:	Orbis
Designer:	Ron Rezek Los Angeles, California
Design Firm:	Ron Rezek Lighting and Furniture Los Angeles, California
Materials:	Cast aluminum, brass, and stainless steel

Product:	Cabriolet
Designer:	M. Coronelli and L. De Licio Milan, Italy
Client:	Thunder & Light/Stilnovo Milan, Italy
Materials:	Lacquered metal housing, opal squared xerographied plexiglas, fluorescent circular bulb

Product: Neon Sconce
Designer: Dan Chelsea and Rudi Stern of Let There Be Neon
New York, New York
Client: George Kovacs Lighting, Inc.
New York, New York
Materials: Black enamel, 100 watt halogen

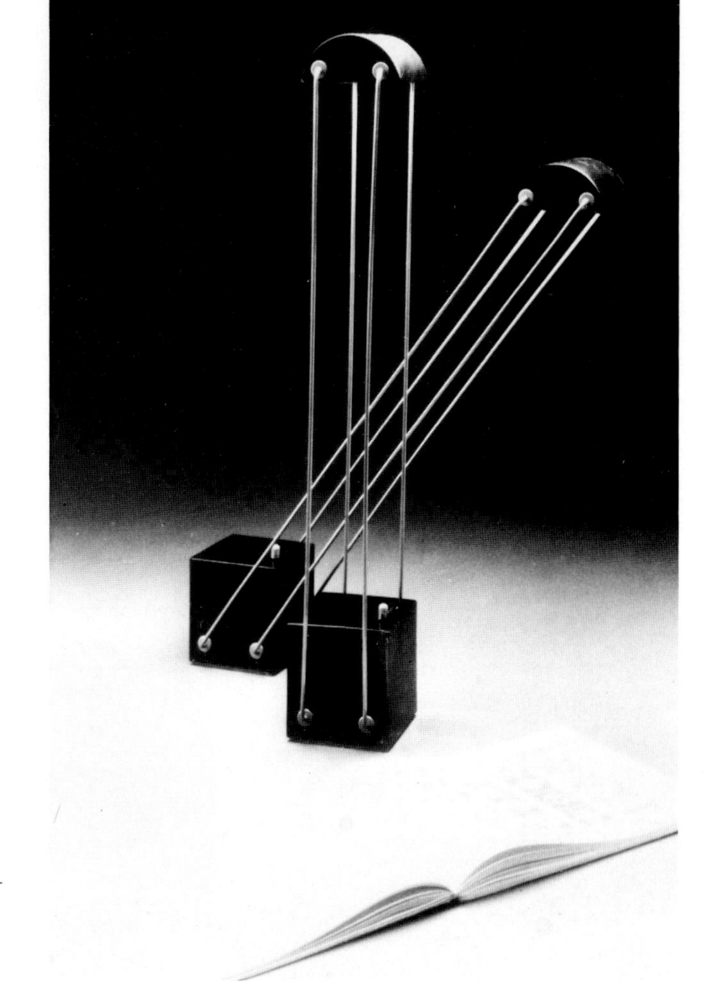

Product: Tokio
Designer: Asahara Sigheaki
Milan, Italy
Client: Thunder & Light/Stilnovo
Milan, Italy
Materials: Makralon reflector and transformer-holder, lacquered metal supports, halogen bulb

Product: Lamps
Designer: Don Ruddy
New York, New York
Design Firm: Furniture Club
New York, New York
Materials: Concrete bases; ribbed, opaline acrylic diffusors; incandescent bulbs

Product: Lamps
Designer: Don Ruddy
New York, New York
Design Firm: Furniture Club
New York, New York
Materials: Concrete bases, rice paper between wire screening, incandescent bulbs

Product:	Gibigiana
Designer:	Achille Castiglioni Milan, Italy
Client:	Flos of Italy Milan, Italy
Courtesy:	Atelier International New York, New York
Awards:	1982 ASID International Product Design award
Materials:	20 watt halogen light source in cylindrical, form-bent metal base

Product:	110 Desk/Table Lamp
Designer:	Ron Rezek Los Angeles, California
Design Firm:	Ron Rezek Lighting and Furniture Los Angeles, California
Materials:	Sheet metal, perforated steel reflector diffusor

Product:	Eubea Lamp
Designer:	Alberto Fraser

Product: Sintesi Professional Task Lamp
Designer: Ernesto Gismondi
Milan, Italy
Client: Artemide, Inc.
New York, New York
Materials: Painted metal; diffusor cup in anodized aluminum with protective black metal grill

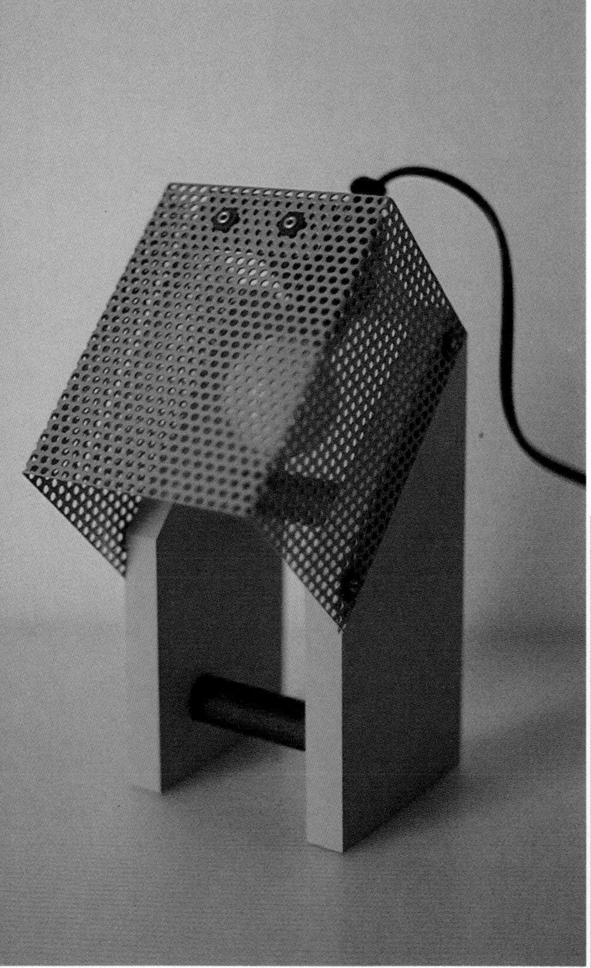

Product: Basic Lamp
Designer: John Willy Campbell
West Yorkshire, England
Materials: Perforated steel and wood, 40 watt Roundlight Bayonet bulb

Product: Directional Lamp
Designer: John Willy Campbell
West Yorkshire, England
Materials: Perforated steel and wood; 40 watt Roundlight Bayonet bulb.

Product: Nastro Lamp
Designer: Alberto Fraser
Client: Stilnovo
Milan, Italy
Materials: Halogen table lamp, multicolored plastic arm, transformer base and reflector, makrolon

Product: Alistro
Designer: Ernesto Gismondi
Milan, Italy
Client: Artemide Inc.
New York, New York
Materials: Swivelling base in metal and molded glassfiber reinforced polyester: adjustable black metal arm; adjustable thermoplastic diffusor; fluorescent bulb

Product: Pausenia
Designer: Ettore Sottsass
Milan, Italy
Client: Artemide Inc.
New York, New York
Materials: Metal, polyurethane resin lacquer finish; chrome diffusor support columns; fluorescent bulb

Product: Lamps
Designer: Alvar Aalto (ca. 1940s and 1950s)
Reproduced by: International Contract Furnishings
New York, New York
Materials: Painted brass and iron

Product: Aton Fluorescent
Designer: Ernesto Gismondi
Milan, Italy
Client: Artemide Inc.
New York, New York
Materials: Lacquered metal with metal base encapsulated in black rubber coating; Fluorescent

Product:	Polifemo
Designer:	Carlo Forcolini London, England
Client:	Artemide Inc. New York, New York
Materials:	Matte black metal, white lacquered metal reflector plate, halogen bulb

Product:	Lamps
Designer:	Alvar Aalto (ca. 1940s and 1950s)
Reproduced by:	International Contract Furnishings New York, New York
Materials:	Painted brass and iron

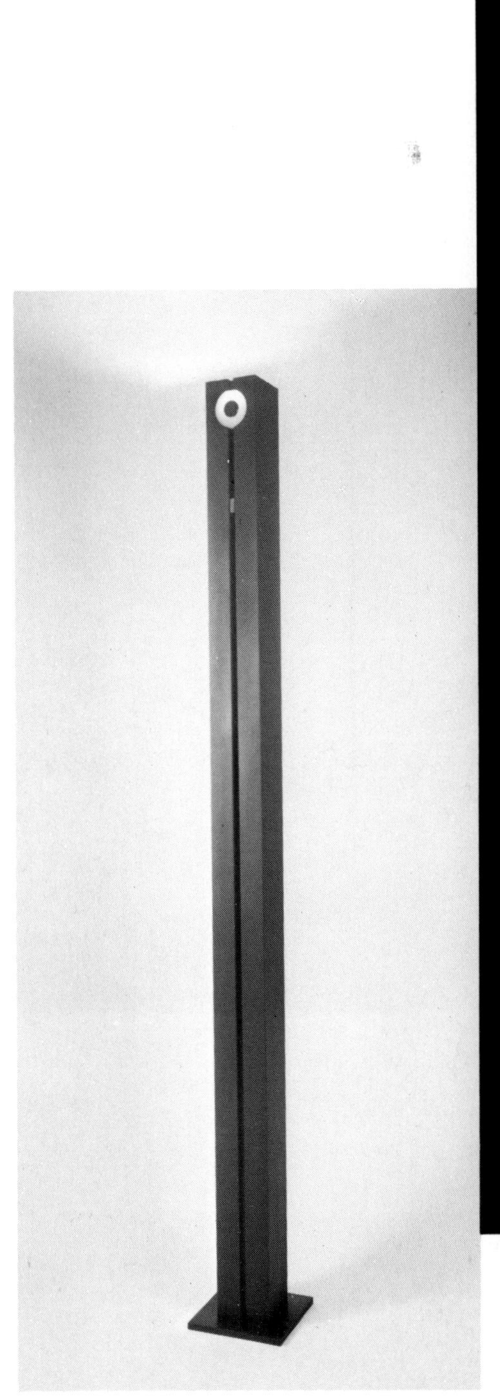

Product: Black Column Torchiere
Designer: Matt Edelstein
New York, New York
Client: George Kovacs Lighting, Inc.
New York, New York
Materials: Aluminum housing with black epoxy matte finish; frosted white flass insert at top; 400 watt halogen with full range sliding dimmer in red

Product: Grand
Designer: Michele de Lucchi
Milan, Italy
Client: Memphis
Milan, Italy
Materials: Metal and plexiglass

Product: Column
Design Firm: Penney & Bernstein,
New York, New York
Client: George Kovacs Lighting, Inc.
New York, New York
Materials: White epoxied aluminum extrusion, 300 watt reflector flood

Product: 610/611 Standing Floor Lights
Designer: Ron Rezek
Los Angeles, California
Design Firm: Ron Rezek Lighting and Furniture
Los Angeles, California
Awards: 1980 IBD Product Design Gold Award
Materials: Sheet metal, perforated steel reflector diffusor

Product: Halogen Torchiere
Designer: Robert Sonneman
New York, New York
Client: George Kovacs Lighting, Inc.
New York, New York
Materials: Solid brass painted black or brass with black details; 400 watt halogen with full range dimmer

Product:	Neon Torch
Designer:	Dan Chelsea and Rudi Stern of Let There Be Neon New York, New York
Client:	George Kovacs Lighting, Inc. New York, New York
Materials:	Anodized aluminum, 400 watt halogen with full range dimmer

Product: Lotek Floor/Table Lamp
Designer: Benno Premsela
Amsterdam, The Netherlands
Client: Eikelenboom Licht and Vorm
Amstel, The Netherlands
Courtesy: Sointu
New York, New York
Materials: Aluminum rods, fabric

Product: Sintesi Track Lamp
Designer: Ernesto Gismondi
Milan, Italy
Client: Artemide Inc.
New York, New York
Materials: Painted metal; diffusor cup in anodized aluminum with protective black metal grill

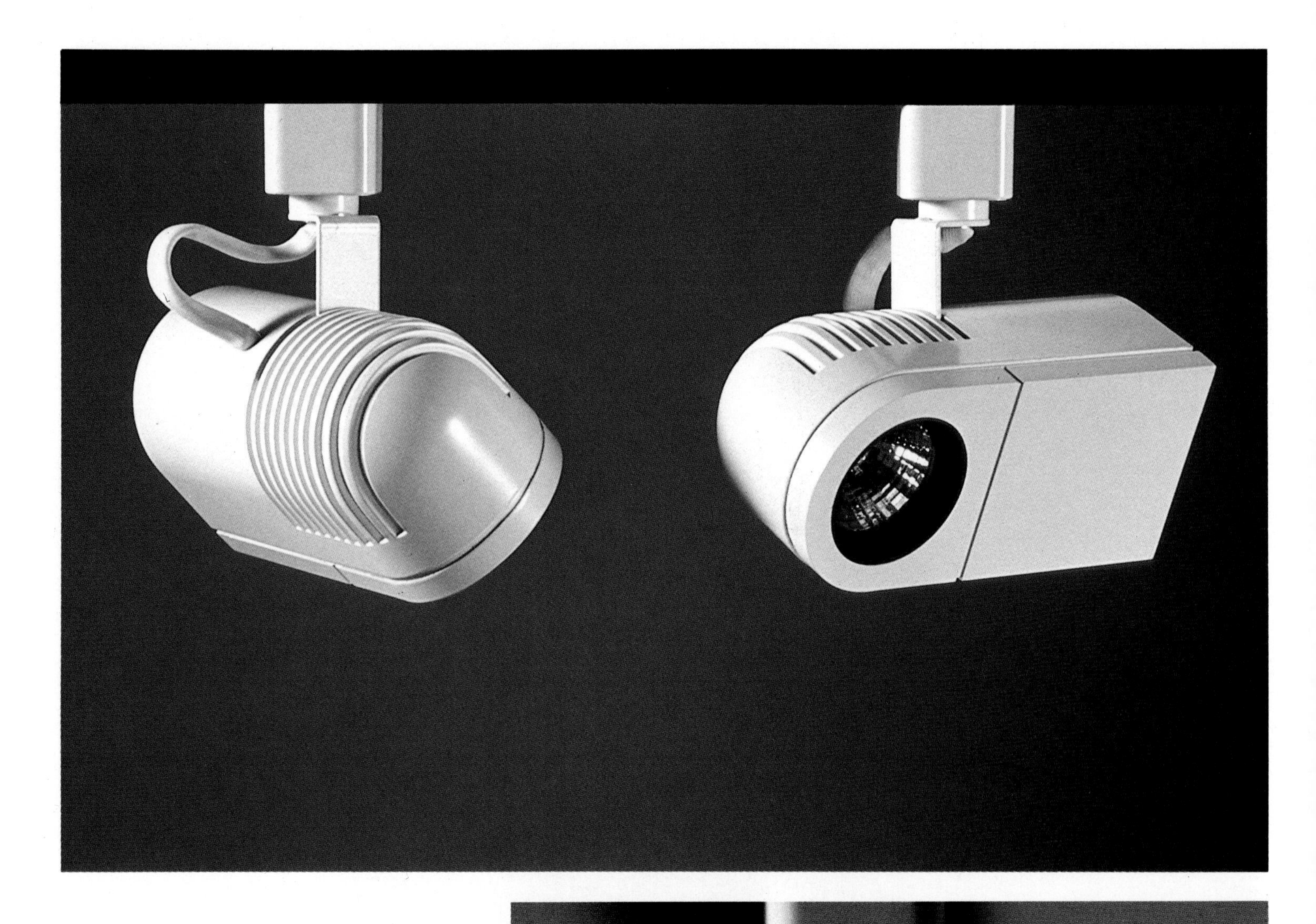

Product:	Halo Track Lighting
Designers:	Scott Roos, Mark Wilson, Ray Kusmer, and Ray Tinley Halo Lighting Division Elk Grove Village, Illinois
Client:	McGraw-Edison Co. Elk Grove Village, Illinois
Awards:	1983 *Industrial Design* magazine Design Review selection
Materials:	Die-cast zinc, semi-gloss white or black finishes

Product:	Oseris Spotlight Range
Designer:	Emilio Ambasz and Giancarlo Piretti New York, New York
Client:	Klaus-Jurgen Maack Ludenscheid, West Germany
Awards:	1983 *Industrial Design* magazine Design Review selection
Materials:	Precast aluminum housing, ceramic sockets, stamped aluminum reflectors, stamped perforated sheet metal heat diffuser

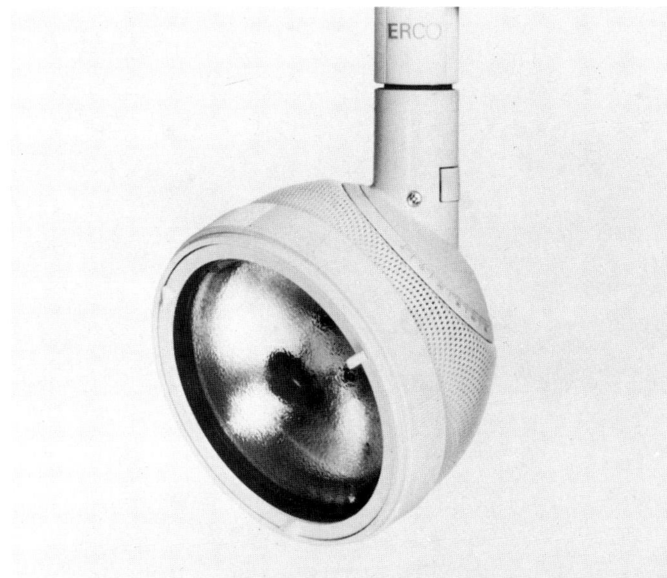

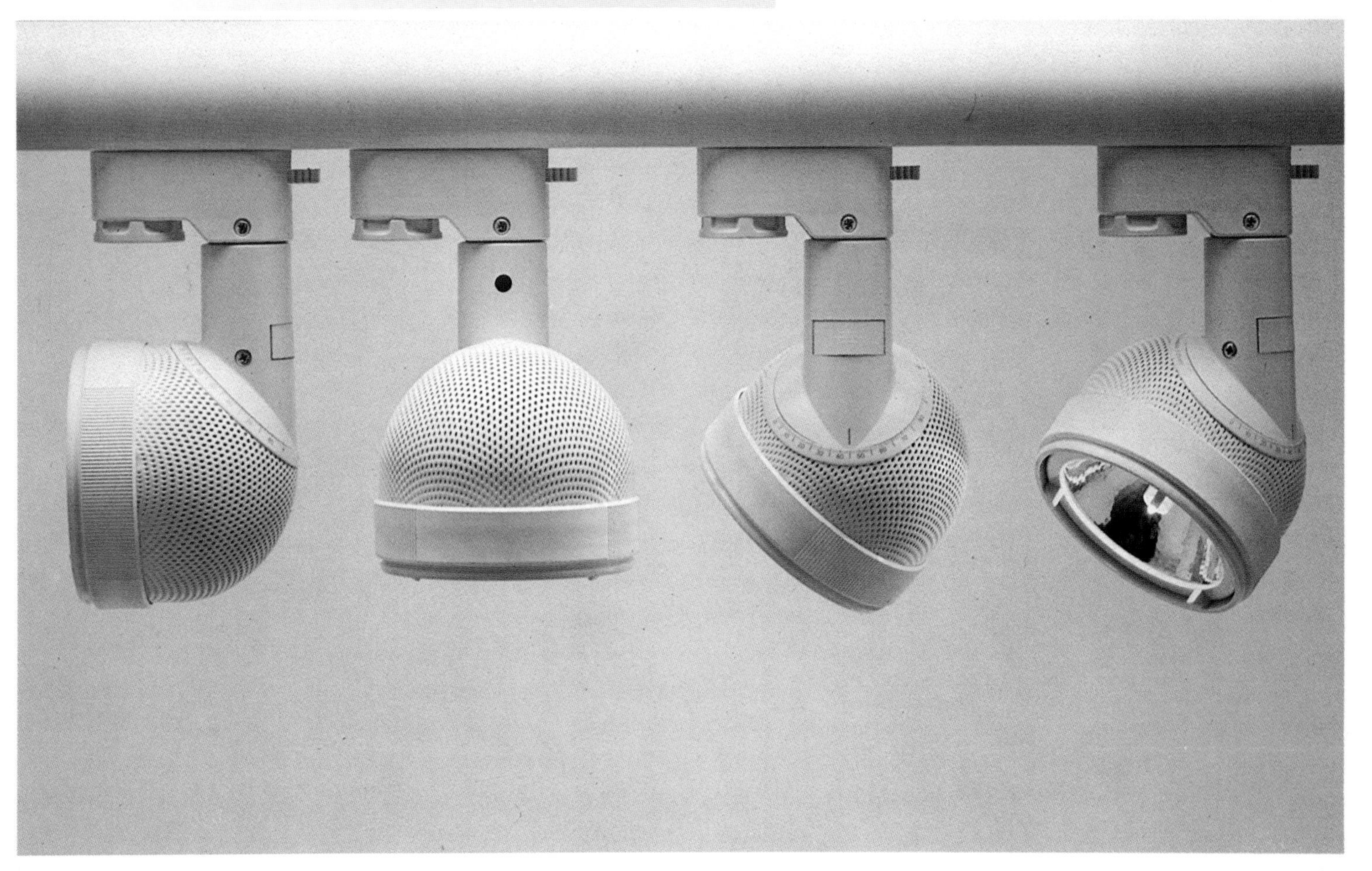

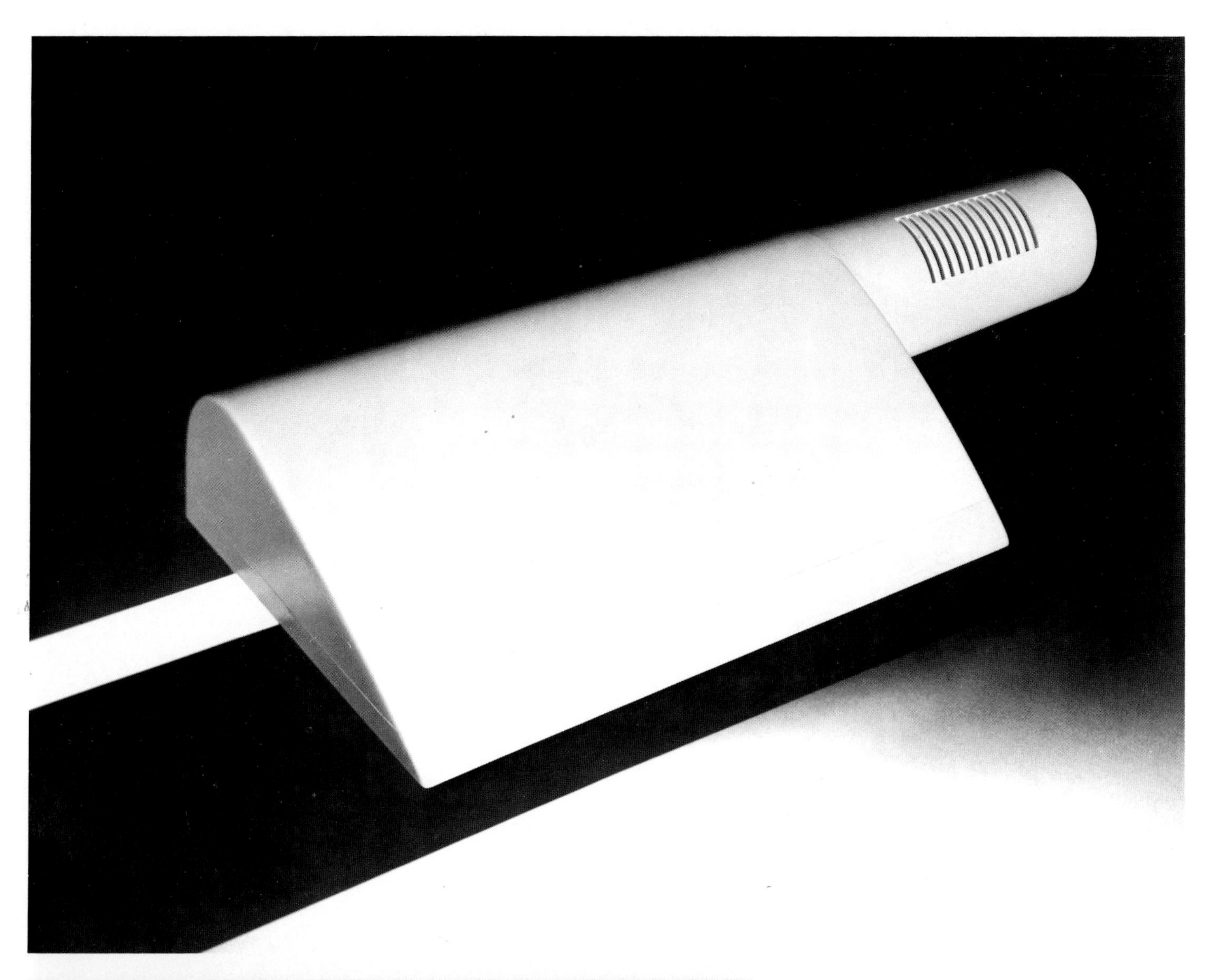

Product: Ledu, IPL 600 assymetrical task light designed for use in offices with computer display terminals
Design Firm: Designspring, Inc.
Westport, Connecticut
Client: Ledu Corporation
Trumbull, Connecticut
Awards: 1983 IDSA Industrial Design Excellence Award
Materials: Injection-molded ABS, Acetal, Durathane, low voltage PL bulb. Reflector and light source system direct light at an angle to the work surface to eliminate glare while maintaining visibility and contrast.

Product: Projectorlite Night Vision Aid
Designer: J. Malan
Bryanston, South Africa
Client: Hazard Equipment (Pty) Ltd.
Bramley, South Africa
Awards: 1982 Shell Design Engineering Product award
Materials: Casing, handgrip, and plastic components molded in fiberglass-reinforced polycarbonate; vulcanized rubber seals; vacuum coated aluminum reflector

Product:	Catseye Emergency Standby Light
Designers:	L. Eaton, A. Livermore, K. Gild, and staff design Durban, South Africa
Design Firm:	Control Logic (Pty) Ltd. Durban, South Africa
Awards:	1982 Shell Design Consumer Product award
Materials:	Extruded aluminum body, standard fluorescent tube

Product:	Lighting Installation
Design Firm:	Penney & Bernstein, New York, New York
Consultant Firm:	CHA Associates New York, New York
Client:	Landor Associates New York, New York
Materials:	Extruded aluminum channels with matte black nonreflective enamel finish; frame integrates wall and ceiling lighting with wiring and cable management systems; vertical lighting, tungsten-halogen wall washer; top component: tungsten-halogen indirect uplights

CHAPTER **4**

Contract and Residential Furnishings

In putting together a collection of contract and residential furnishings a number of questions are raised, the foremost of which is whether the two categories can be combined, or if it is more appropriate to treat them separately. The former course was chosen simply because so many of the products could have fallen into either category. An expansive mahogany table, for example, would be as suitable in a corporate conference room as in the dining area of a loft. Although office systems have yet to find applications in the home, and many of these pieces really can serve only in one area or the other, a great many others can adapt to either. What this signals, perhaps, is a shift in the way we perceive our homes and the places in which we work. While more and more people are devising ways in which they can work out of their homes, those who stay in the office appear to be looking for ways to make them look warmer and more comfortable—more, that is, like home. So it comes as no surprise that so many of these furnishings are serviceable in either place.

Another noticeable trend in recent furnishings is the homage much of it pays to history. Eileen Gray, Rennie Mackintosh, Le Corbusier, and a great many others have had their work reintroduced to the marketplace. While these designs, most now between 50 and 80 years old, are scarcely contemporary, they have been included in this chapter because they also represent current sensibilities. Whether it is for reasons of "vulgar commercialism" in the words of one designer, or a more thoughtful contemplation of form, designers are searching history for forms that may bear repetition. They are doing so in numbers high enough to indicate that it is neither aberrant nor anachronistic, but rather a substantive part of the way we view contemporary furnishings. Not to recognize this would also be failing to recognize the full scope of contemporary design sensibility.

What is also noticeable in this selection of furnishings is the number of pieces designed by architects. This is not exactly a novel idea. Most of the furniture in the Modern Movement was designed by architects who wanted the furnishings in a building's interior to correspond to the exterior form in its sense of proportion, attention to material and technological process, and the like. The inclusion of pieces designed by architects in this chapter, however, may need some justification. Although most of the furniture is conspicuous for its attention to form, material, and craft, it is less outstanding for its attention to the demands of production. Often, the designs are exercises in aesthetics that pay little heed to production costs, manufacturing processes, and the demands of the market. Still, these pieces are valuable reference points; often, it is the artist or designer working in one area of design who discovers what can be done in another. That designer, because he or she is less informed, may also have a less constrained vision of what possibilities exist. This is certainly not to say that industrial designers should look to architects for new ways to make tables, but simply that innovations can, and often do, come from outside a particular field. As often as an industrial designer designs a well-proportioned table that can be manufactured and marketed at a reasonable cost, an architect can come up with a form which, though it may be impractical, uses a refreshing color, has a whimsical curve, or uses a new laminate or building material that a designer more accustomed to making tables might not think to use.

What may signal these pieces most dramatically is their attention to—or captivation by—color and ornament. Though these elements may severely try the patience of any manufacturer adventuresome enough to take them on, the fact that they have entered the vocabulary of contemporary furnishings is enough in itself. Their message is implicit: Whether they go so far as to respond to "the need for our emotional participation in our environment" in the words of one Memphis advocate, or whether they are visual delights that are meant to give only pleasure, they attest finally that appearance need not reflect the pragmatc purpose of furniture.

They also ask that we shift our conventional attitudes toward furniture. Rather than look at a chair in terms of how it can be accommodated stylistically in a furnished room, or how it maintains some aesthetic continuity, these pieces ask that we accept them more individually. A chair, like a painting or ceramic vessel, can provoke, startle, and amaze as well as provide a place to sit.

Which brings us to the point of comfort. Many of the chairs included in this chapter, for example, rather shamelessly discourage the idea of sitting in them. Although this, at first, may seem to defeat the whole purpose of making a chair, it actually, on second thought, does not. What these designers seem to recognize is that although a great many chairs are indeed meant to be sat on, there are plenty of others which provide different services; while some chairs are meant to accommodate the human body as it works or eats, others wind up being a place to lace your shoes or hang a sweater. Since these chairs do not need to support the human frame for hours on end, they might as well look good doing what little they do.

If visual provocation rather than comfort marks much of what is new in residential furnishings, office systems pay all the closer attention to ergonomics. Witness the proportions of Neils Diffrient's Helena chair. The natural arc of the human body is nearly perfectly translated in the arc of the chair's back. Explains Diffrient: "Over the range of recline, you'll have a differential of only an eighth of an inch, which you won't even feel." The Helena Chair is only one example of these fine-tuned anthropometrics that acknowledge the need for office furniture to comfortably support the seated human frame for long stretches of time.

Another example is the Dorsal seating system by Emilio Ambasz and Giancarlo Piretti, which changes its configuration automatically whenever the user changes his or her position. Studies indicated that the frequent movement from a forward leaning position to a relaxed reclining position contributed to a healthy spine by encouraging the diffusion of nutrients in intervertebral disks. The backrest of the Dorsal seating system provides the user with support in any position, and thereby encourages this movement.

Ergonomics in the office, of course, is not limited to seating systems. Entire office systems can be—and are—designed to address ergonomics. In question particularly is how the office can sensibly integrate the rapid growth of electronic equipment with user need and comfort. The Activity Center Modules from Harvey Probber include glare-free laminated desk tops with adjustable height and tilt, acoustical privacy screens, and wire management ducts. The net effect is one of a personalized office space that accommodates individual postures and physiques, and adapts to diverse functions. Likewise, Ergodata by Precision Mfg. Co. is an office system which is based upon the integration of electronic data into the office. Again, wire and cable management channels and adjustable work surfaces are the keys that make it so flexible. While the ergonomics of designing a comfortable chair presents enough problems, designing a whole system geared toward both electronic equipment and its user is even more ambitious, and is perhaps what most distinguishes the office system of the 1980s. What this assures us, too, is that the whimsical attention paid to form in residential furnishings is balanced by the scrupulous attention to function found in office furnishings.

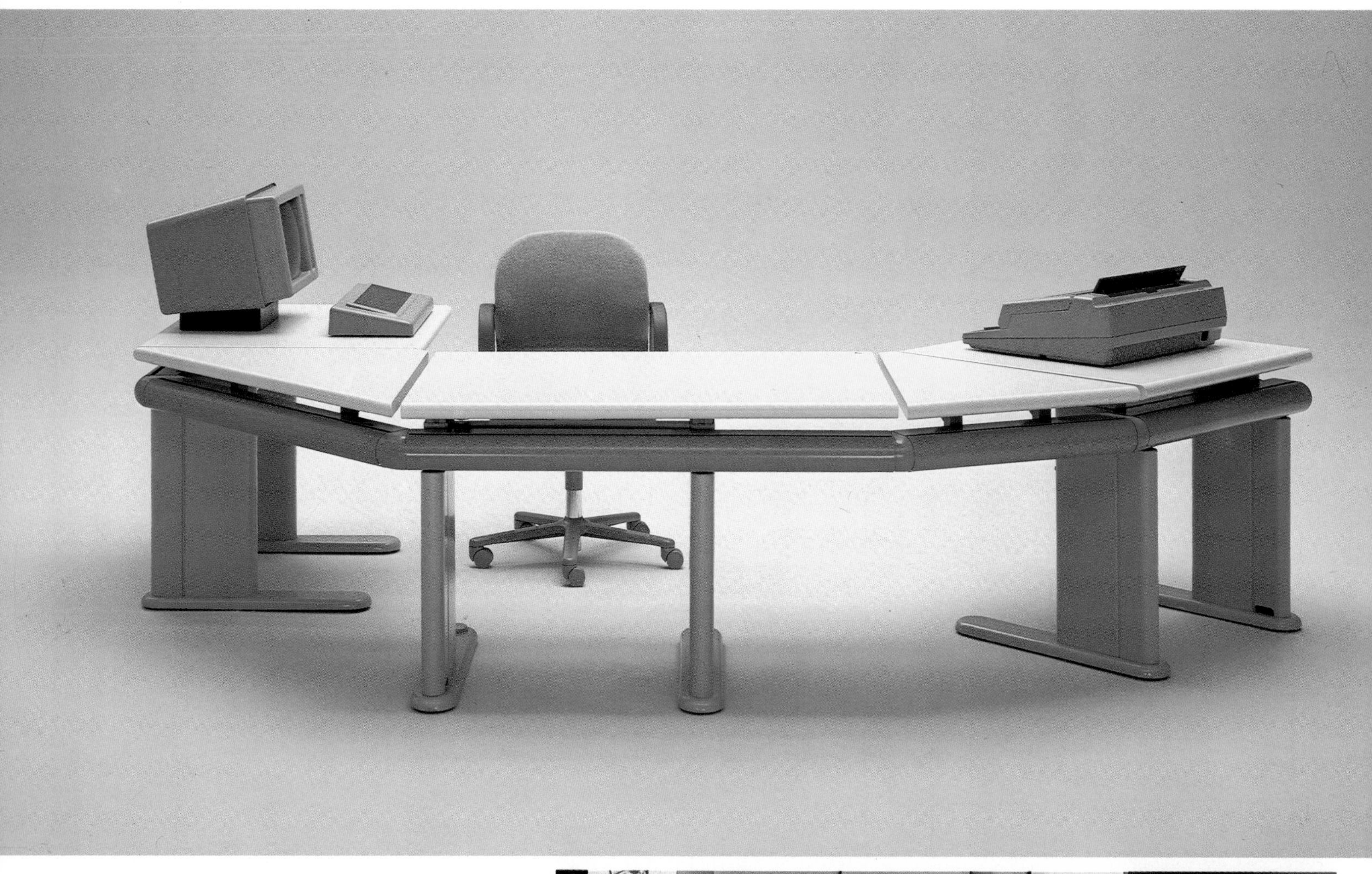

Product:	Dunbar S/4 Series
Designers:	Jack Dunbar, Lydia dePolo, and Steven Brooks New York, New York
Design firm:	dePolo/Dunbar, New York, New York
Client:	Dunbar, Berne, Indiana
Awards:	1982 IBD Product Design silver award
Materials:	Reconstituted woods, plastic laminates, coordinated fabrics

Product:	ACM ergonomically designed component system
Designer:	Karl Dittert, Giessen, West Germany
Client:	Probber/Voko, New York, New York
Awards:	1983 IBD Product Design silver award 1983 Roscoe Product Design award
Materials:	Glare-free postformed laminate desk tops and steel desk bases

Product:	Action Office Color, Fabric, Finish, and Texture System
Designer:	Clino Castelli, Milan, Italy
Client:	Herman Miller, Inc., Zeeland, Michigan
Materials:	Include coordinated silkweaves, flannels, polyknits, perforated vinyls, and brushknits

Product:	COM System (Continuum)
Designers:	Francesco Frascaroli and C. Biondi, Bologna, Italy
Clients:	C.O.M., Bologna, Italy Krueger, Inc., Green Bay, Wisconsin
Awards:	1983 IBD Product Design gold award
Materials:	Legs: modular cast aluminum; Surfaces: high density particleboard with veneer or laminate finish; Connecting beams: steel

Product:	COM System (Interval)
Designers:	Francesco Frascaroli and C. Biondi, Bologna, Italy
Client:	C.O.M., Bologna, Italy Krueger, Inc., Green Bay, Wisconsin
Awards:	1983 IBD Product Design gold award
Materials:	Legs: modular cast aluminum; Surfaces: high density particleboard with veneer or laminate finish; Connecting beams: steel

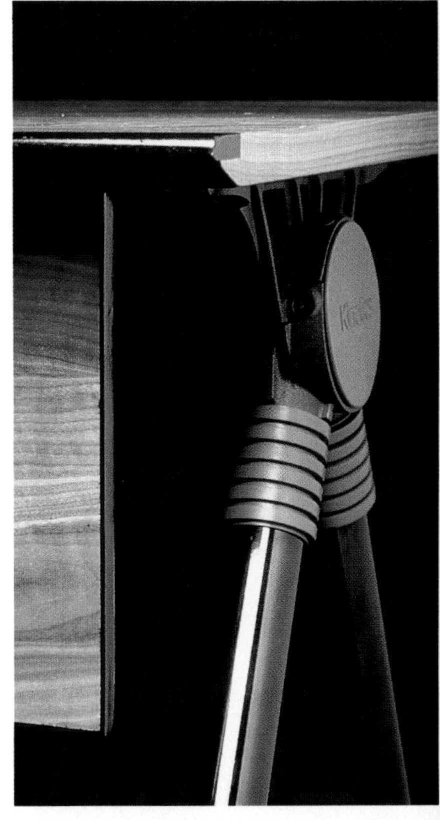

Product: Reception Desk System with segregated channels for electrical circuits and communications cables
Designers: Paolo Favaretto, Padova, Italy and Jim Hayward, Rexdale, Ontario, Canada
Client: Kinetics, Rexdale, Ontario, Canada
Materials: Frame: steel; Surface: veneer or plastic laminate

Product: Executive Desk System with segregated channels for electrical circuits and communications cables
Designers: Paolo Favaretto, Padova, Italy and Jim Hayward, Rexdale, Ontario, Canada
Client: Kinetics, Rexdale, Ontario, Canada
Materials: Frame: steel; Surface: veneer or plastic laminate

Product: General Office Desk System with segregated channels for electrical circuits and communications cables
Designers: Paolo Favaretto, Padova, Italy and Jim Hayward, Rexdale, Ontario, Canada
Client: Kinetics, Rexdale, Ontario, Canada
Materials: Frame: steel: Surface: veneer or plastic laminate

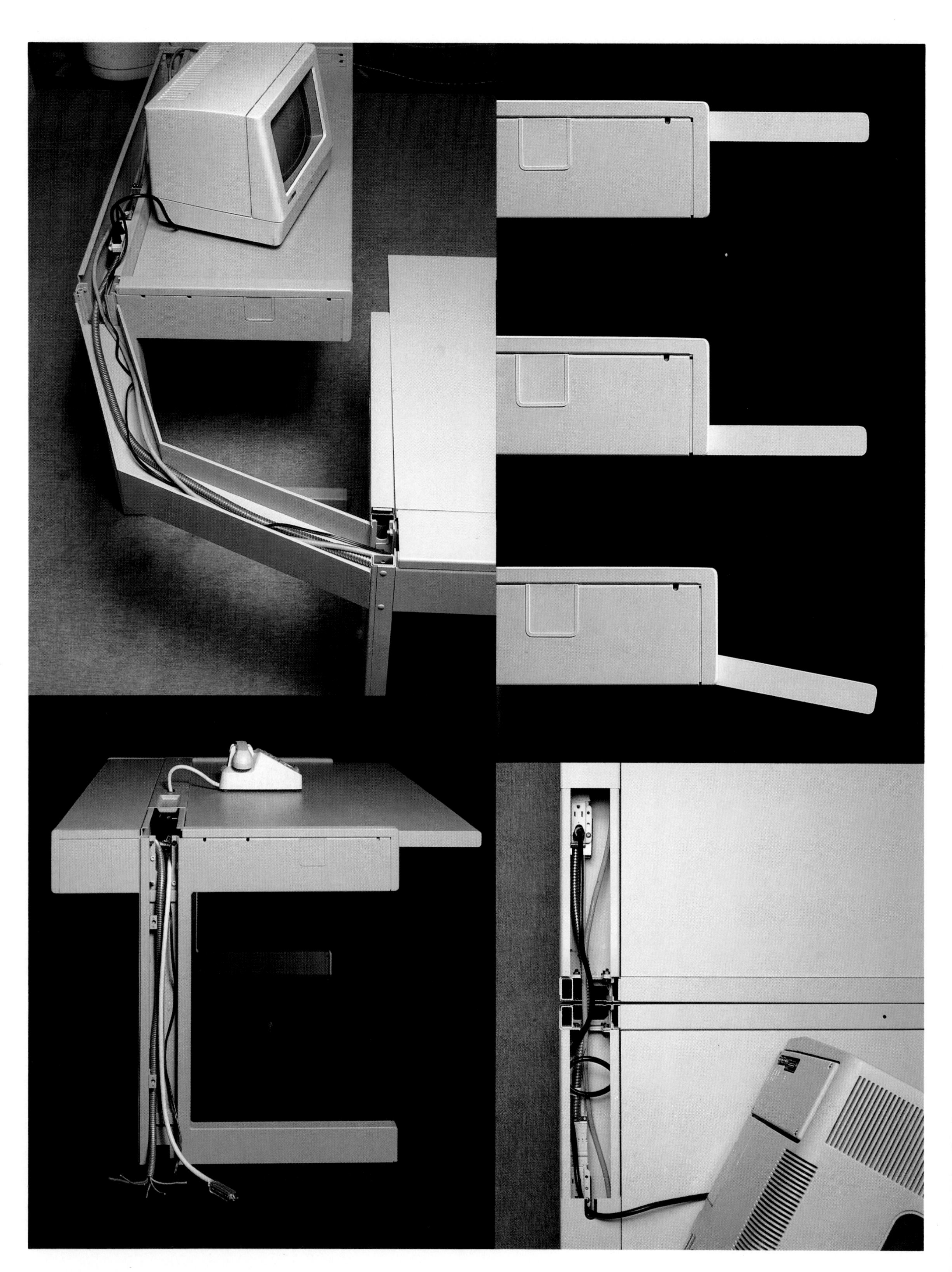

Product:	Ergodata Office System with adjustable work surfaces, acoustic panels, organizational channels for power and communication cables
Designer:	Urs Bachmann, Zurich, Switzerland
Client:	Precision Manufacturing Company Montreal, Quebec, Canada
Awards:	1983 IBD Product Design honorable mention
Materials:	Surfaces: plastic laminate; Core: wood; Frame: steel

Product: Dolmen office furnishings
Designer: Gino Gamborini, Bologna, Italy
Client: Castelli Furniture Inc., Bohemia, New York
Awards: 1982 IBD Product Design silver award
Materials: Panel: chipboard veneered with tay wood; solid wood edging inserts and detailing in Miura leather

Product: The Burdick Group
Designer: Bruce Burdick, San Francisco, California
Client: Herman Miller, Zeeland, Michigan
Awards: 1981 IDSA Industrial Design Excellence Award
Materials: Beams: polished aluminum; Surfaces: wood, marble, glass, or laminate

Product: Data Entry Work Station
Designer: Bruce Hannah, Cold Spring, New York
Client: Knoll International, New York, New York
Materials: ABS plastic; steel frame; wood drawers

Product: Gwathmey Siegel Desk
Designer: Charles Gwathmey and Richard Siegel New York, New York
Client: Knoll International, New York, New York
Materials: Drawers and compartment shelves: vinyl covered fiberboard; hardwood drawer pulls; Finishes: Techgrain or mahogany veneer

Product:	Alpha Office System
Design Firm:	Frogdesign, Campbell, California
Client:	Konis & Neurath, Karben, West Germany
Awards:	Stüttgart Design Center "Excellence of Design" award
Materials:	Steel; particleboard with formica finish; aircraft technology operates lift-and-tilt mechanism

Product:	SK-7 Desk
Designer:	William Sklaroff, Philadelphia, Pennsylvania
Design Firm:	William Sklaroff Design Associates Philadelphia, Pennsylvania
Client:	Gunlocke, Waylande, New York
Materials:	Rift cut natural oak

Product:	Pinstripe Office Furniture
Designers:	Walker/Group, Randy Stultz and Mark Zeff New York, New York
Design firm:	Walker/Group Inc. New York, New York
Client:	ICF, New York, New York
Awards:	1983 IBD Product Design gold award
Materials:	Mahogany, cherry, walnut, ash, or oak

Product:	Trading Desk
Design Firm:	Penney & Bernstein New York, New York
Client:	Salomon Brothers Inc. New York, New York
Manufacturer:	Spec Built Carlstadt, N.J.
Materials:	Walnut; Nevamar plastic laminate; aluminum; glass
Photo Credit:	Dan Cornish New York, New York

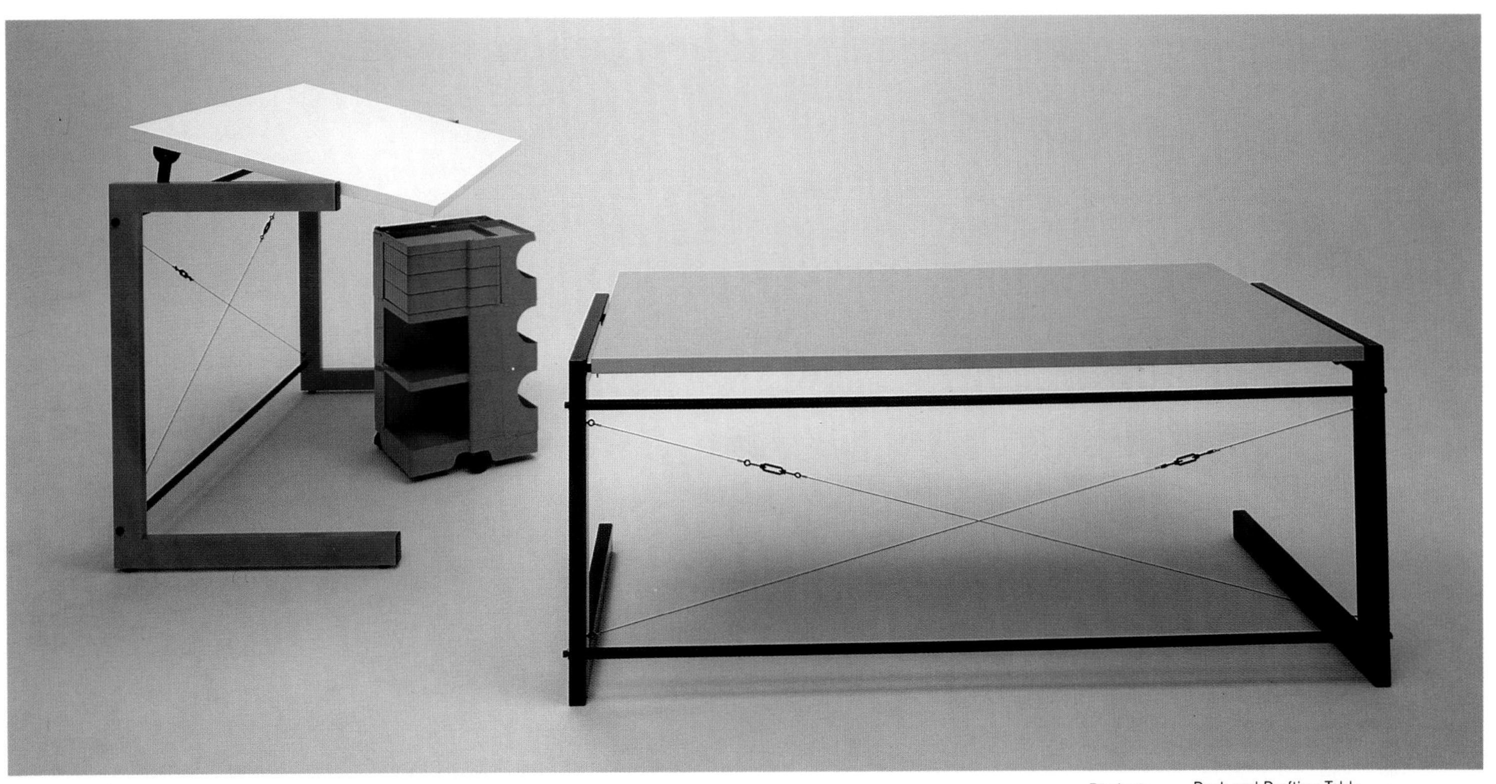

Product: Desk and Drafting Table
Designer: Ron Rezek, Los Angeles, California
Design Firm: Ron Rezek Lighting + Furniture
Los Angeles, California
Awards: 1981 IBD Product Design gold award
Materials: Legs: heavy steel with plastic coating; crossed tension cables and tubular compression rods; Surfaces: Formica Colorcore laminate

Product: Balans Activ 6000 and adjustable Activ Working Table
Designer: Hans Chr. Mengshoel and Svein Gusrud, Oslo, Norway
Client: HAG USA Inc., Chicago, Illinois subsidiary of HAG A/F, Oslo, Norway
Materials: Frame: tubular stainless steel with upholstered polyurethane cushions; Tabletop: wood or plastic laminate

Product: Balans Vital 6035
Designer: Hans Chr. Mengshoel and Peter Opsvik, Oslo, Norway
Client: HAG USA Inc., Chicago, Illinois subsidiary of HAG A/F, Oslo, Norway
Materials: Base: aluminum; pneumatic lift; Seat: foam/fabric covered plywood

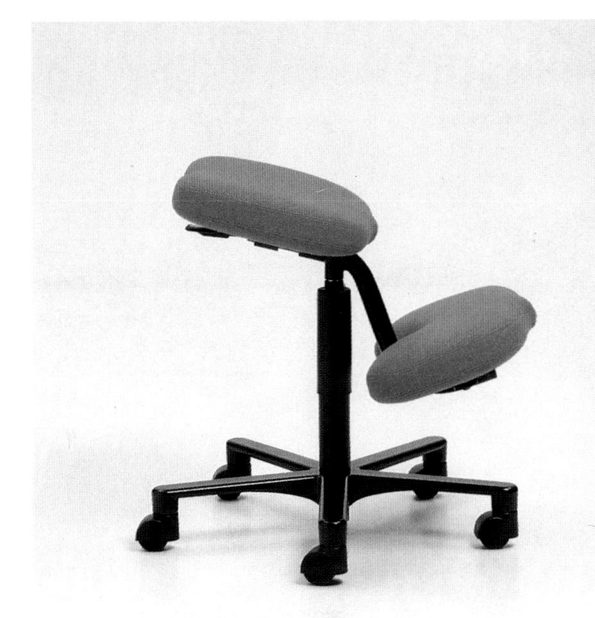

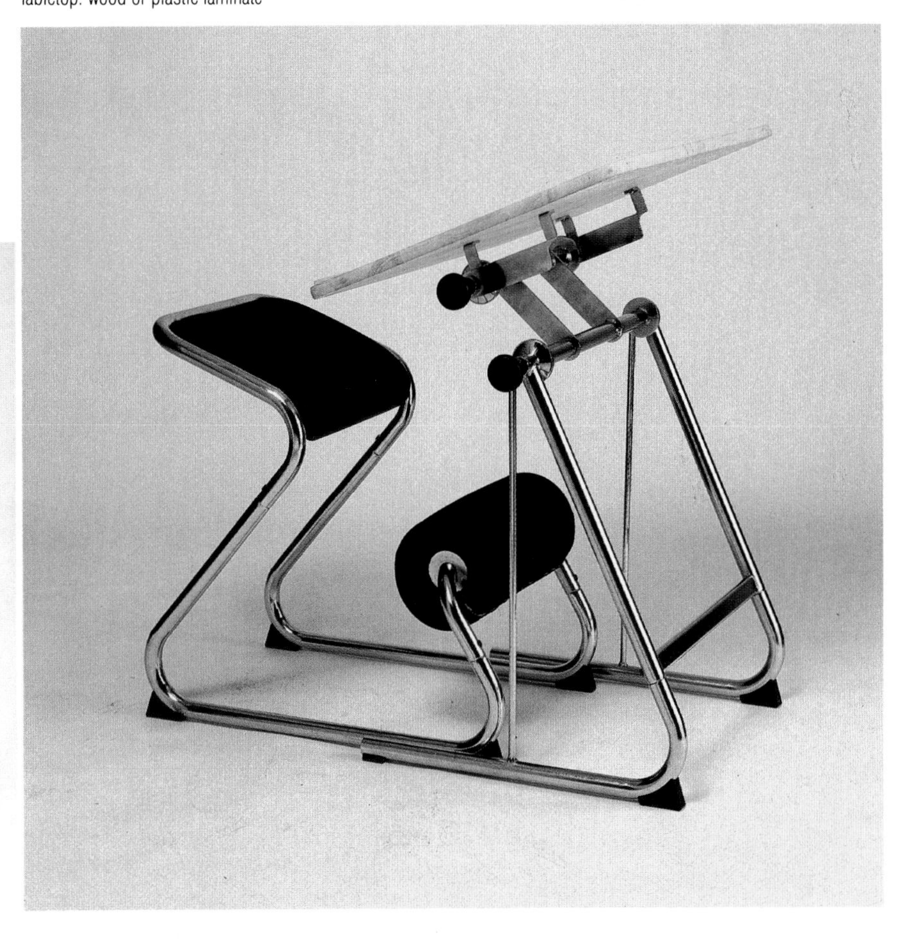

Product: Stephens Office Seating/Management Chair
Designer: Bill Stephens, East Greenville, Pennsylvania
Client: Knoll International, New York, New York
Materials: Steel base; Finish: fused polyester; Upholstery: fabric, vinyl, or leather

Product: Stephens Office Seating/Executive Chair
Designer: Bill Stephens, East Greenville, Pennsylvania
Client: Knoll International, New York, New York
Materials: Steel base; Finish: fused polyester; Upholstery: fabric, vinyl, or leather

Product: Stephens Office Seating/Task Chair
Designer: Bill Stephens, East Greenville, Pennsylvania
Client: Knoll International, New York, New York
Materials: Steel base; Finish: fused polyester; Upholstery: fabric, vinyl, or leather

Product: Stephens Office Seating
Designer: Bill Stephens, East Greenville, Pennsylvania
Client: Knoll International, New York, New York
Materials: Steel base; Finish: fused polyester; Upholstery: fabric, vinyl, or leather

Product: Kevi™ Chair
(L-R): high-back management chair; sled base/pull-up chair; operational stool; secretarial chair; low-back management chair
Client: Herman Miller, Inc., Zeeland, Michigan
Materials: Shell: polystyrene; Armrests: polyurethane foam; Finish: polished aluminum or black umber epoxy base

Product: Dorsal™ Seating Range
Designer: Emilio Ambasz and Giancarlo Piretti
New York, New York
Materials: Seats and backrests: injection molded thermoplastic; Hinge: steel with integral hinge mechanism; Legs: seam welded oval tubular steel components

Product: Sirkus Chair
Designer: Yrjo Kukkapuro, Helsinki, Finland
Client: Avarte Oy, Helsinki, Finland
Materials: Tubular steel and form-pressed veneer with detachable upholstery

Product: Helena Chair
Designer: Neils Diffrient, Ridgefield, Connecticut
Client: Sunar/Hauserman, Norwalk, Connecticut
Awards: 1982 *Industrial Design* magazine Design Review selection 1982 *Time* magazine Chair of the Year
Materials: Seat: hardwood shell with metal brackets, molded polyurethane foam upholstery; Arm supports: one-inch heavy gage steel tubing, chrome plated; Base: 5-pronged, 22-inch-diameter glass reinforced Rynite with black gloss finish

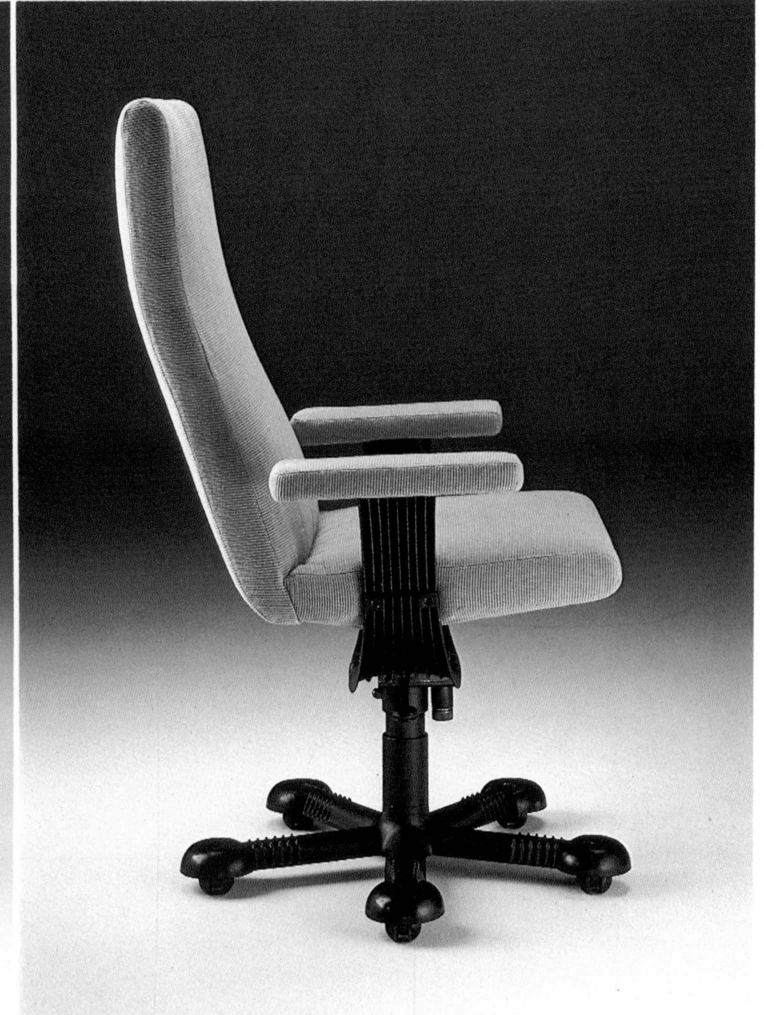

Product: Kinetics Business Seating
Designer: Paolo Favaretto, Padova, Italy
Client: Kinetics, Rexdale, Ontario, Canada
Materials: Polymer frame and custom upholstery

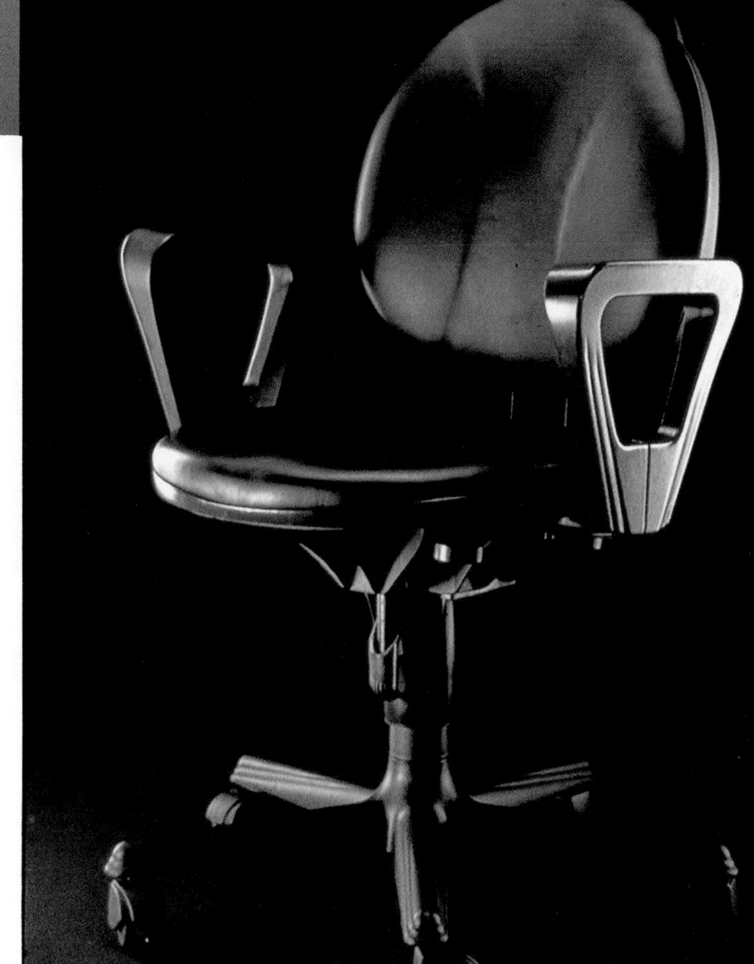

Product: Mix executive chairs and Master desks and conference tables
Designers: Afra and Tobia Scarpa, Milan, Italy
Client: Stendig International Inc., New York, New York
Awards: 1982 IBD Product Design honorable mention
Materials: Executive chairs: cast aluminum with black enamel matte finish; on casters or glides; fabric or leather upholstery. Desks and tables: Oak, walnut, or rosewood veneer surface wrapped with black leather; die-cast aluminum leg exterior with interior face of molded structural black plastic

Product:	Wire Structure Combinations
Designer:	Paul Haigh, New York, New York
Design Firm:	Haigh Architecture & Design, New York, New York
Awards:	1983 *Progressive Architecture* magazine furniture competition award
Materials:	Base structure: wire-form; Shells and arms: thin-wall die cast zinc; Upholstery: neoprene

Product:	Public Chair Prototype
Designer:	Toshiroh Ikegami, Osaka, Japan
Awards:	1st International Design Competition Osaka, Japan

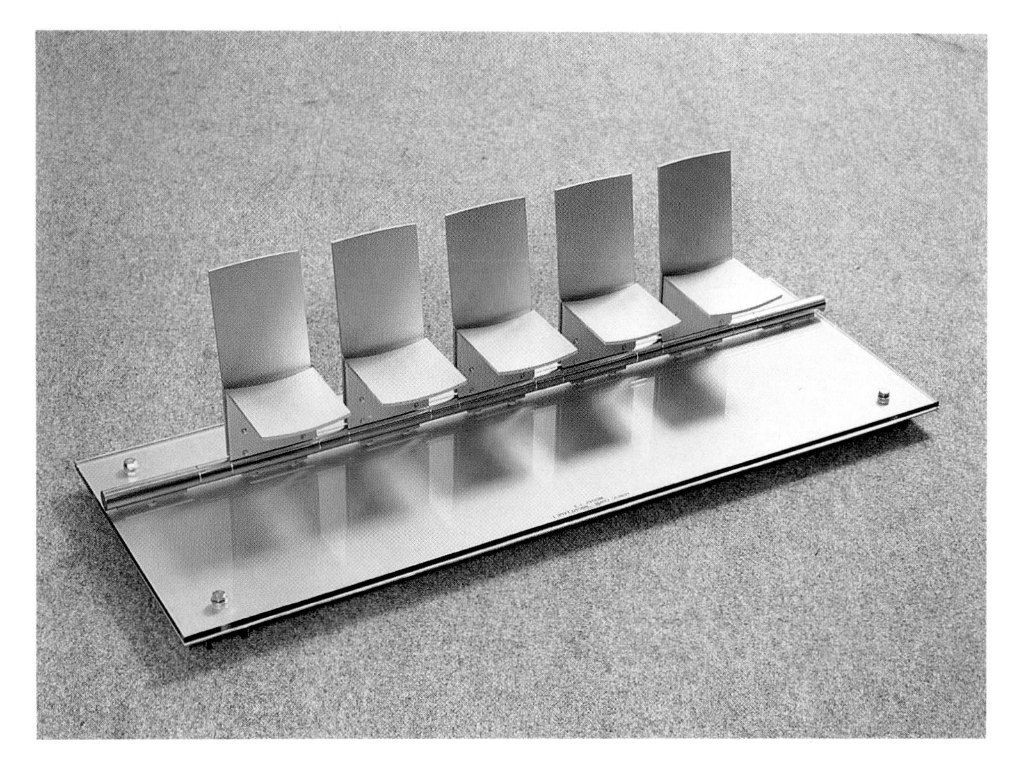

Product:	Flip-Seat™
Designer:	David Goodwin, London, England
Design Firm:	Goodwin-Wheeler Associates, London, England
Client:	Fixtures Furniture, Kansas City, Missouri
Awards:	1983 IBD Product Design gold award 1980 Design Council design Award
Materials:	Seat: molded polypropylene; Frame: heavy gauge steel finished in epoxy

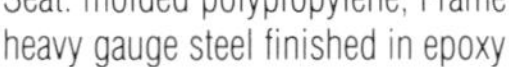

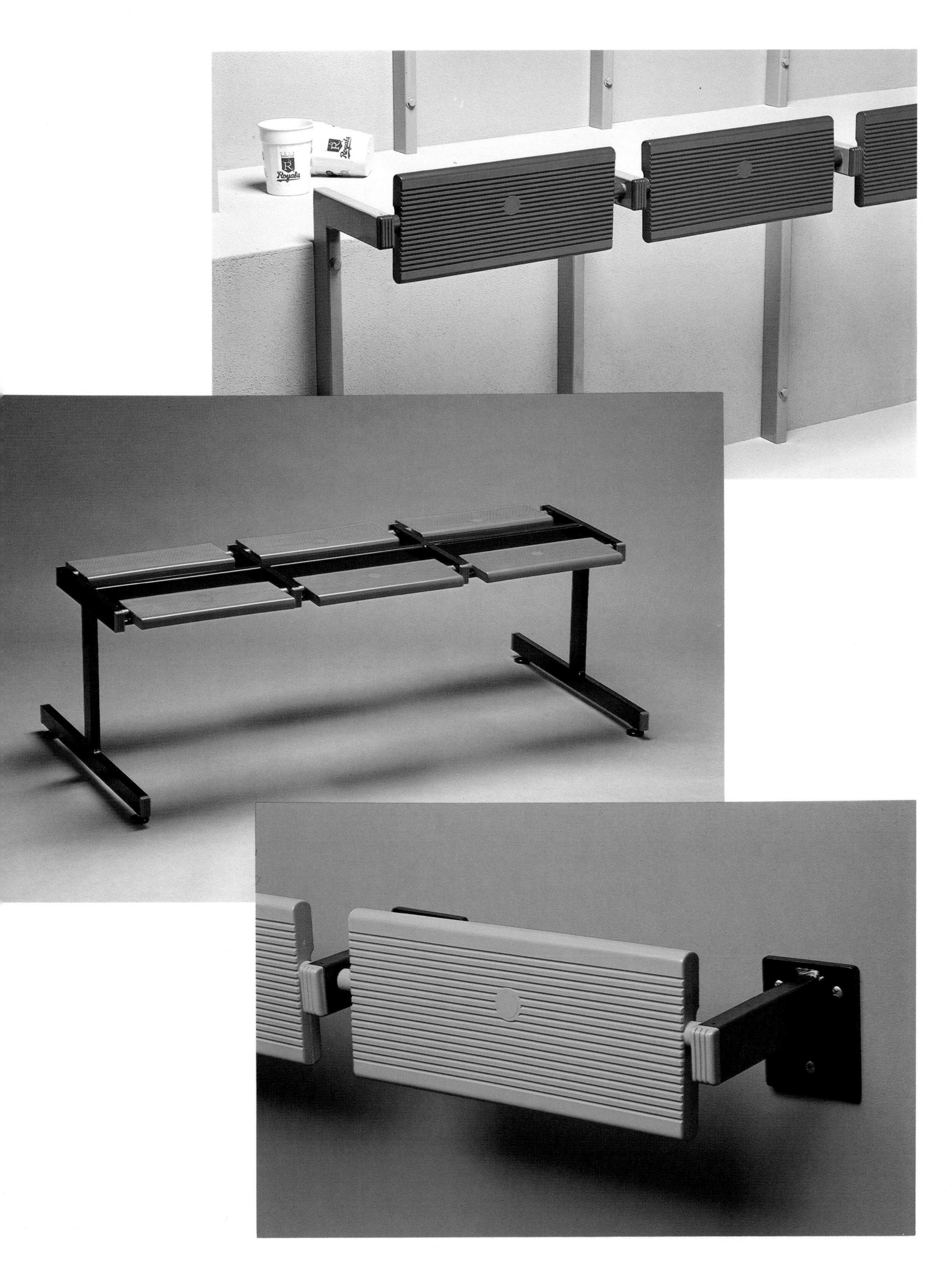
Royals
Royals

Product: Gina Chair
Designer: Bernd Makulik, Dudinghausen, West Germany
Client: Stendig International Inc., New York, New York
Awards: 1982 IBD Product Design gold award
1982 Roscoe award
Materials: Frame: solid beech with fabric, vinyl, or leather upholstery

Product: Courthouse Chair
Designer: Walker/Group, Randy Stultz
New York, New York
Design Firm: Walker/Group Inc., New York, New York
Client: The Gunlocke Company, Wayland, New York
Materials: Walnut

Product: Klassik & Klasse
Designer: Ernst Dettinger, Munich, West Germany
Client: Jack Lenor Larsen, New York, New York
Awards: 1981 Roscoe Product Design award
Materials: Solid beech

Product:	Uni Chair
Designer:	Werther Toffoloni, Udine, Italy
Client:	Atelier International, New York, New York
Awards:	1983 *Industrial Design* magazine Design Review selection
Materials:	Beechwood frame and legs; aluminum inserts secure legs to hardwood seat

Product:	The Grid Chair
Designer:	Ward Bennett, New York, New York
Design Firm:	Ward Bennett Designs for Brickell Associates New York, New York
Materials:	Kiln-dried solid ash with mahogany, natural oiled ash, black wax finishes and upholstered seat cushion

Product:	Andover Chair
Designer:	Davis Allen, New York, New York
Design Firm:	Skidmore, Owings, and Merrill New York, New York
Client:	Stendig International, Inc., New York, New York
Materials:	Solid beechwood with fabric or leather upholstery

Product:	MartinStoll Collection/G
Manufacturer:	Martin Stoll, Tiengen, West Germany
Awards:	National "Gute Form 1983" Prize
Materials:	Solid beechwood and steel

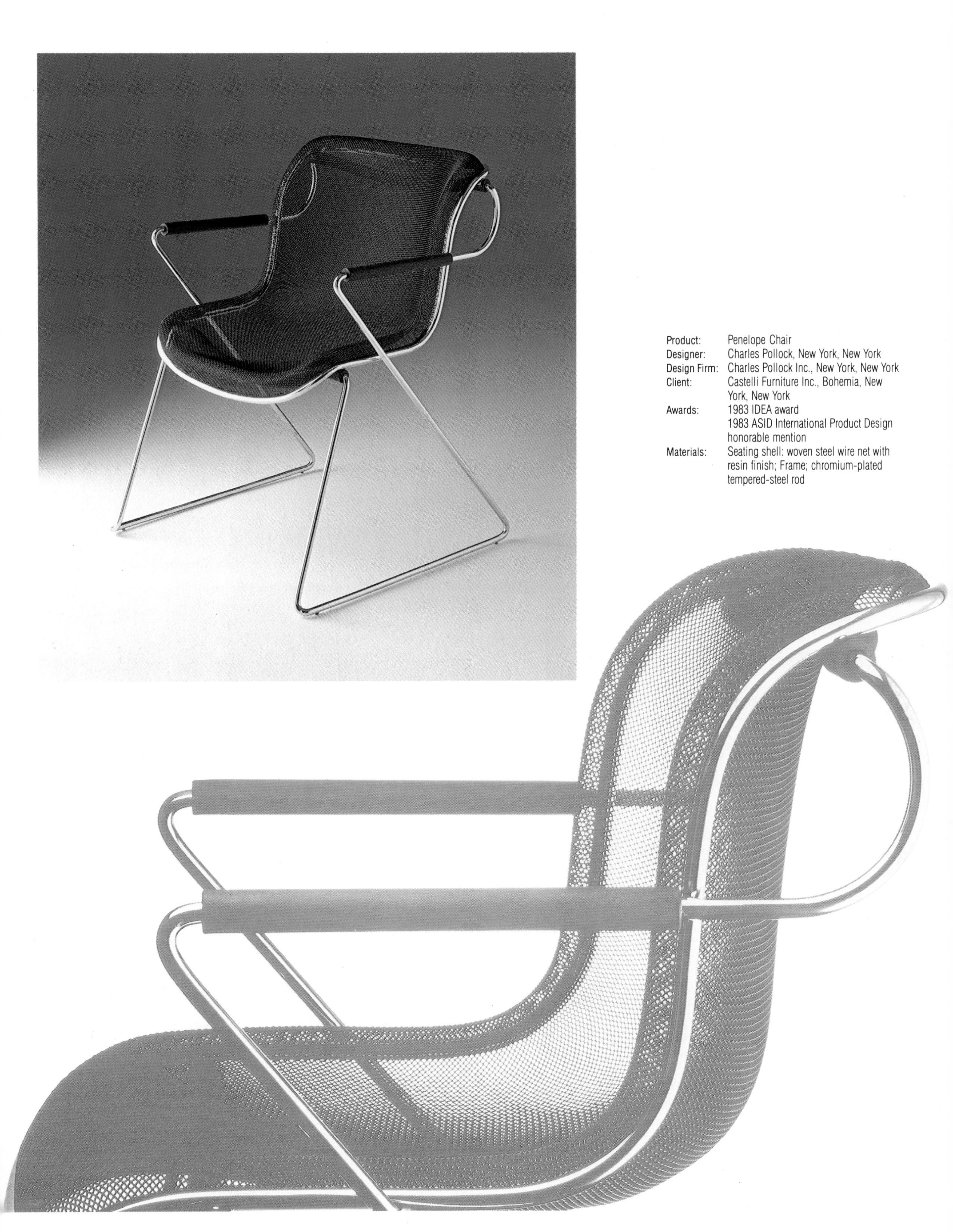

Product:	Penelope Chair
Designer:	Charles Pollock, New York, New York
Design Firm:	Charles Pollock Inc., New York, New York
Client:	Castelli Furniture Inc., Bohemia, New York, New York
Awards:	1983 IDEA award 1983 ASID International Product Design honorable mention
Materials:	Seating shell: woven steel wire net with resin finish; Frame; chromium-plated tempered-steel rod

Product:	Schultz Chair
Designer:	Richard Schultz, East Greenville, Pennsylvania
Client:	Knoll International, New York, New York
Materials:	Steel tubing with painted or chrome finish; leather, vinyl, or fabric upholstery

Product:	Kita Collection
Designer:	Toshiyuki Kita, Milan, Italy
Client:	Stendig International Inc., New York, New York
Awards:	1983 IBD Product Design honorable mention
Materials:	Beech and ash with matte polyurethane finish

Product:	Stacking Chair
Designer:	Keith Muller, Mark Campbell Toronto, Ontario, Canada
Design Firm:	Keith Muller Limited Toronto, Ontario, Canada/Washington D.C.
Client:	Simò-Dow Manufacturing Limited Calgary, Alberta, Canada
Awards:	1983 *Industrial Design* magazine Design Review selection
Materials:	Seat: plywood with shell-natural varnish finish on maple: or birch-faced plywood; Upholstered seat; 1½-inch steel tubing and ⅜-inch C.R.S. rods

Product:	Everychair Series
Designer:	Peter Danko, Alexandria, Virginia
Design Firm:	Peter Danko & Associates, Alexandria, Virginia
Awards:	1983 IDEA award
Materials:	Integrated molded plywood shell formed by laminating plastic into molded plywood

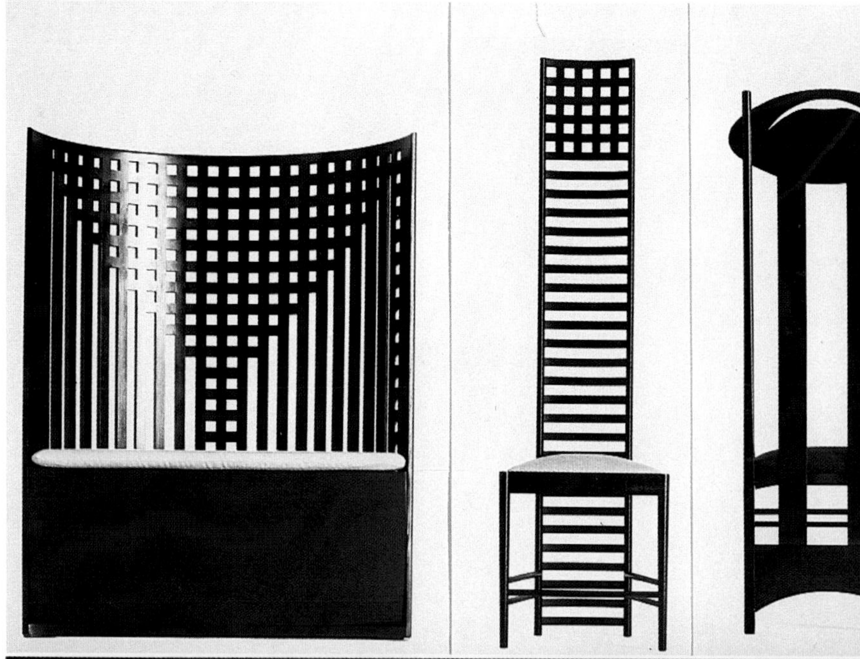

Product: Willow 1 (1904), Hill House 1 (1902), Argyle (1897)
Designer: Charles R. Mackintosh
Reproduced by: Cassina of Italy, Milan, Italy
Materials: Willow 1: ebonized ashwood frame with upholstered frame; Hill House 1: ebonized ashwood frame with upholstered frame; Argyle: ebonized ashwood frame with upholstered frame

Product: LC 2 (1928)
Designer: Le Corbusier
Reproduced by: Cassina of Italy, Milan, Italy
Materials: Frame: polished chrome plated or enamel steel; Padding: polyurethane and Dacron; Upholstery: fabric or leather

Product: D.S., 2–4 (1918)
Designer: Charles R. Mackinstosh
Reproduced by: Cassina of Italy, Milan, Italy
Materials: Ebonized ashwood frame inlaid with mother of pearl and sea grass seat

Product: Transat Armchair (1927)
Designer: Eileen Gray
Produced by: Ecart International, Paris, France
Distributed by: Furniture of the Twentieth Century Inc. New York, New York
Materials: Frame: lacquered wood; Details: nickel plated; Upholstery: black leather

Product: Willow 2 (1904)
Designer: Charles R. Mackintosh
Reproduced by: Cassina of Italy, Milan, Italy
Materials: Ebonized ashwood frame inlaid with mother of pearl

Product: Zig-Zag (1934)
Designer: Gerrit T. Rietveld
Reproduced by: Cassina of Italy, Milan, Italy
Materials: Unfinished or finished elm

Product: Red and Blue (1918)
Designer: Gerrit T. Rietveld
Reproduced by: Cassina of Italy, Milan, Italy
Materials: Beechwood frame with lacquer finish on seat

Product:	Lounge Chair (1929)
Designer:	Rene Herbst
Produced by:	Escart International, Paris, France
Distributed by:	Furniture of the Twentieth Century, Inc. New York, New York
Materials:	Finish: Nickel-plated steel or semi-matte black

Product:	Inna
Designer:	Pennti Hakala, Helsinki, Finland
Client:	Inno, Inno-tuote Oy, Helsinki, Finland
Materials:	Plywood seat; linen cloth; canvas or leather cushions

Product:	Carlos Riart Chair
Designer:	Carlos Riart, Barcelona, Spain
Client:	Knoll International, New York, New York
Awards:	1983 Roscoe Product Design award
Materials:	Ebony with Brazilian amaranthe with mother-of-pearl inlays or holly wood with ebony inlays

Product:	Slim Chair
Designer:	Johan Huldt, Sweden
Client:	Conran's, New York, New York
Materials:	Removable cotton cover on tubular steel frame and polypropelene sling

Product: Botta Chair
Designer: Mario Botta, Lugano, Switzerland
Client: ICF, New York, New York
Materials: Frame: steel tubing with silver or black epoxy; Seat: perforated steel with silver or black epoxy; Back: self-skinned expanded charcoal polyurethane

Product: Steamer Collection
Designer: Thomas Lamb, Uxbridge, Ontario, Canada
Client: Ambiant Systems Ltd., Toronto, Ontario, Canada
Awards: Permanent Design Study Collection of the
Museum of Modern Art
1982 Roscoe Award
1983 IBD Product Design gold award
Materials: Veneer: Canadian maple coated in clear lacquer or black stain

Product: Conference Table
Designers: Shane Kennedy and Don Ruddy, New York, New York
Design Firm: Furniture Club, New York, New York
Materials: Top: cinder-gray concrete; Legs: sulphur yellow concrete

Product: Coffee table
Designer: Shane Kennedy and Don Ruddy, New York, New York
Design Firm: Furniture Club, New York, New York
Materials: Legs: terra cotta; Top: glass

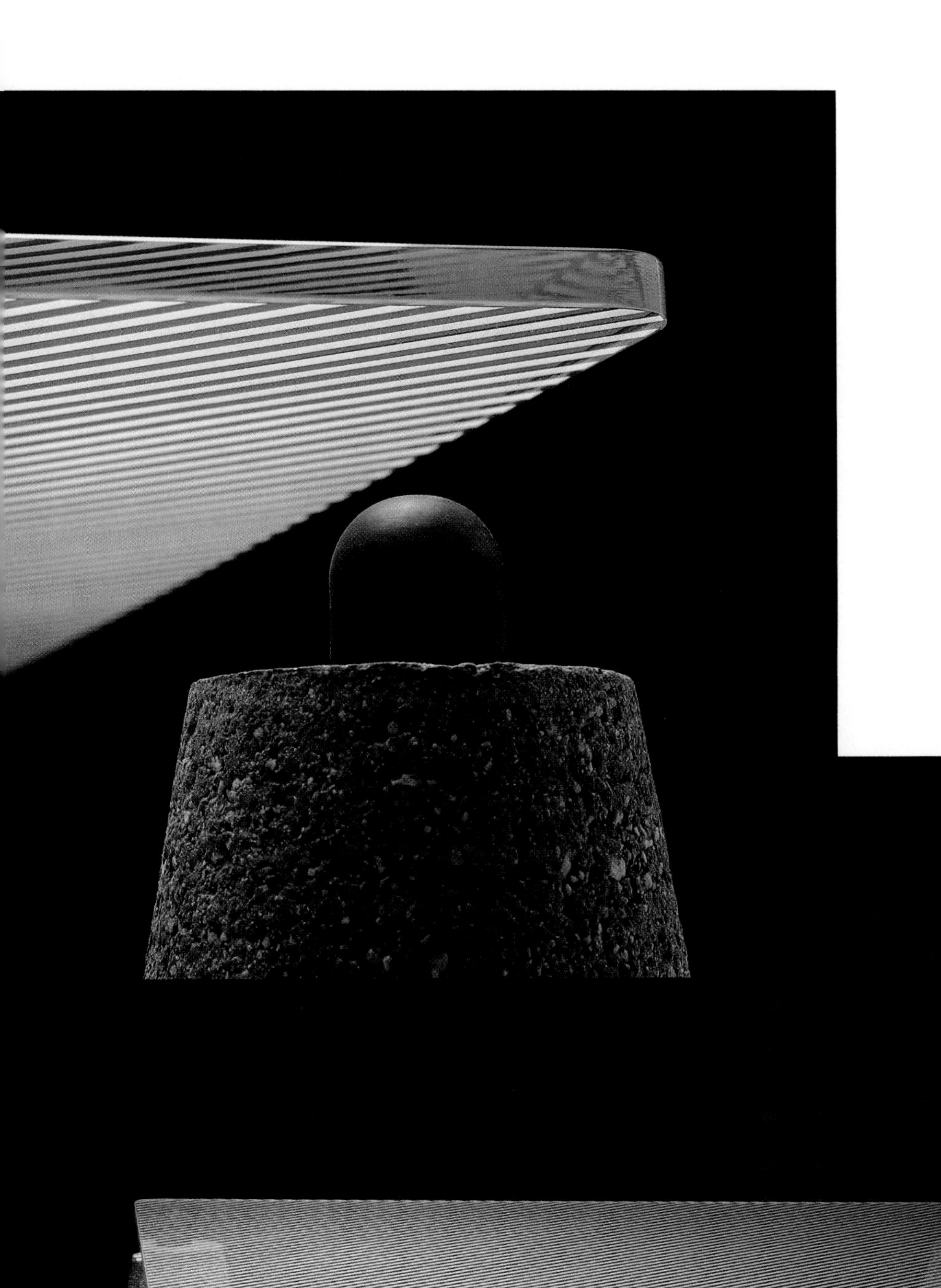

Product:	Pina Gamba
Designer:	Geoffrey Frost, Topanga, California
Design Firm:	Frost Design, Topanga, California
Awards:	1983 IBD Product Design silver award
Materials:	Concrete and glass

Product:	45 Table
Client:	Krueger, Green Bay, Wisconsin, USA
Awards:	1982 IBD Product Design honorable mention 1983 *Industrial Design* magazine Design Review selection
Materials:	Beveled table edges: solid oak for oak veneer and plastic laminate surfaces; solid hardwood for matching walnut, teak, or rosewood veneers. Legs, aprons, corner joiners: powder-coated epoxy finishes

Product: Bistro Table
Designer: Anna Castelli Ferrieri, Milan, Italy
Client: Beylerian Limited, New York, New York
Awards: 1983 *Industrial Design* magazine Design Review selection
Materials: Thermoplastic, techopolymer surface finished with antiscratch paint; polypropylene legs

Product:	Hospital bedside table
Designers:	Keith Muller, Anne Carlyle, Mark Campbell Toronto, Ontario, Canada
Design Firm:	Keith Muller Limited Toronto, Ontario, Canada/Washington, D.C., USA
Client:	Alberta Children's Hospital Calgary, Alberta, Canada
Awards:	1983 *Industrial Design* magazine Design Review selection
Materials:	Wood with natural finish; plastic tray top

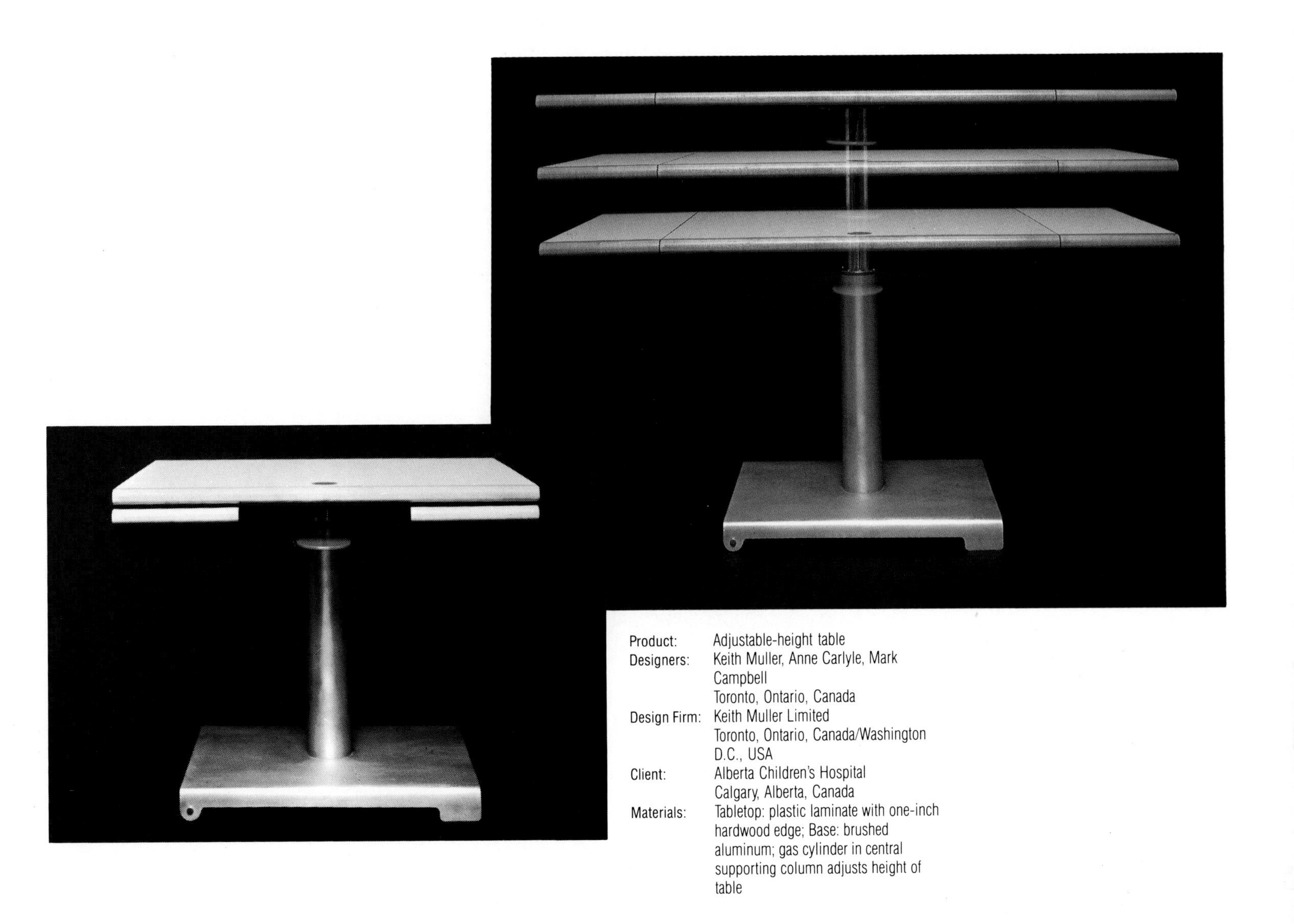

Product: Adjustable-height table
Designers: Keith Muller, Anne Carlyle, Mark Campbell
Toronto, Ontario, Canada
Design Firm: Keith Muller Limited
Toronto, Ontario, Canada/Washington D.C., USA
Client: Alberta Children's Hospital
Calgary, Alberta, Canada
Materials: Tabletop: plastic laminate with one-inch hardwood edge; Base: brushed aluminum; gas cylinder in central supporting column adjusts height of table

Product: Lucia Mercer Collection
Designer: Lucia Mercer, Vermont
Client: Knoll International, New York, New York
Materials: Granite

Product:	Natasa Child's Cot
Designer:	Biba Bertok, Ljubljana, Yugoslavia
Client:	Slovenijales, Ljubljana, Yugoslavia
Materials:	Wood and linen

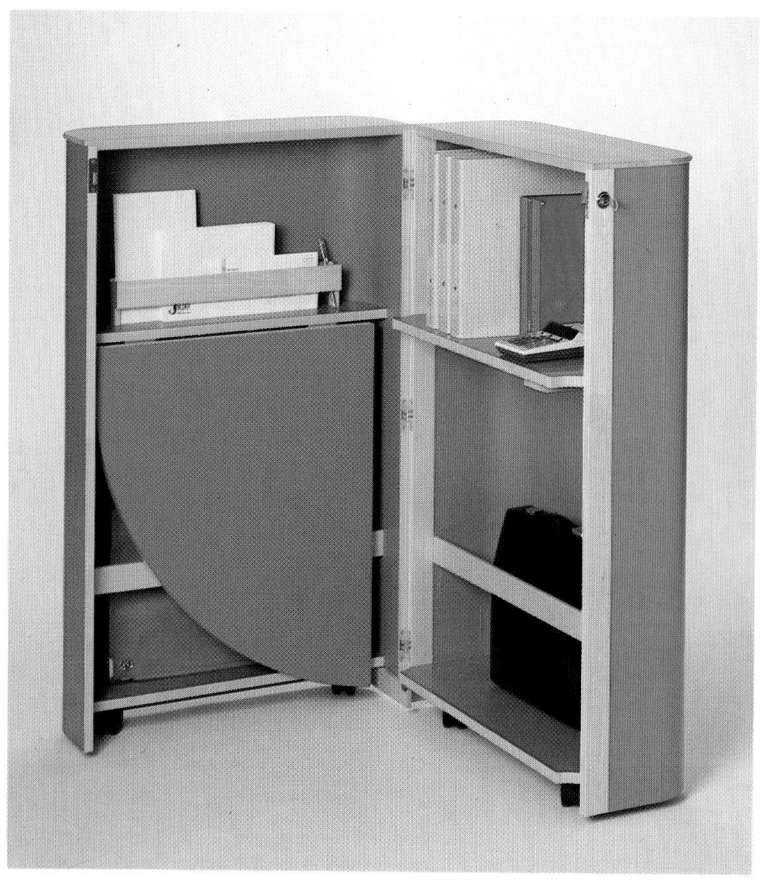

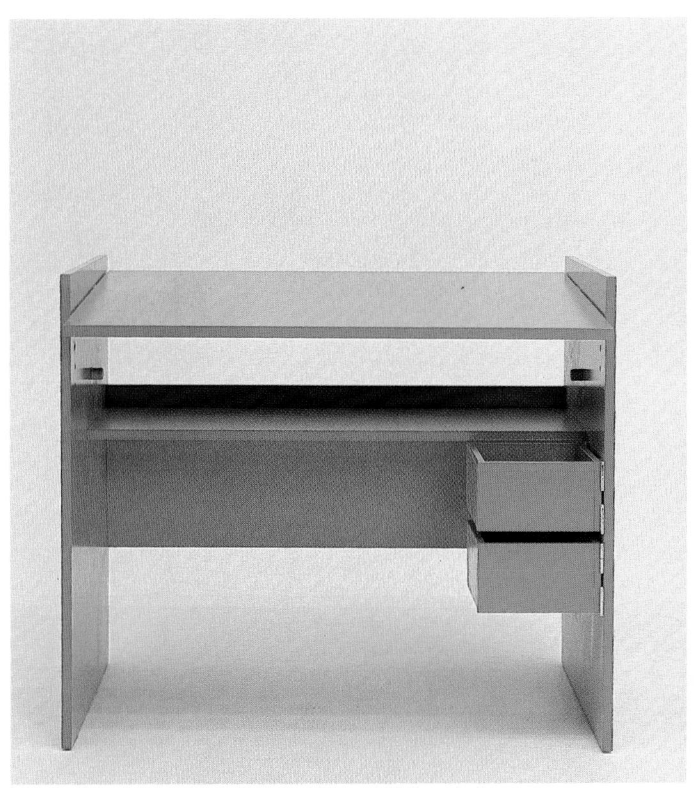

Product:	"Lili" writing table and blackboard for children
Designer:	Tatjana Coloni, Ljubljana, Yugoslavia
Client:	Slovenijales, Ljubljana, Yugoslavia
Materials:	Lacquered chipboard

Product:	Post-Box
Design Firm:	Lindau & Lindekrantz, Ahus, Sweden
Client:	Ahmans i Ahus AB, Ahus, Sweden
Materials:	Plastic laminate; detail in solid maple

Product: Temple Chair
Designers: Jim Lewis and Clark Ellefson Columbus, South Carolina
Design Firm: Lewis & Clark, Columbus, South Carolina
Designed for: Colorcore™ "Surface & Ornament" Competition I and Exhibition
Awards: "Surface & Ornament" Competition, 1st Prize winner
Materials: Colorcore™ surfacing material by Formica Corporation

Product: Strata
Designer: Brian Faucheux, Metairie, Louisiana
Designed for: Colorcore™ "Surface & Ornament" Competition I and Exhibition
Awards: "Surface & Ornament" Competition, 3rd Prize Winner
Materials: Colorcore™ surfacing material by Formica Corporation

Product: Neapolitan
Designer: Lee Payne, Atlanta, Georgia
Design Firm: Lee Payne Associates, Inc., Atlanta, Georgia
Designed for: Colorcore™ "Surface & Ornament" Competition I and Exhibition
Awards: "Surface & Ornament" Competition, 2nd Prize winner
Materials: Colorcore™ surfacing material by Formica Corporation

Product: Classical Cabinet
Designer: Paul Chiasson, New York, New York
Design Firm: P.D.C.C. Studio, New York, New York
Designed for: Colorcore™ "Surface & Ornament" Competition I and Exhibition
Awards: "Surface & Ornament" Competition, 4th Prize Winner
Materials: Colorcore™ surfacing material by Formica Corporation

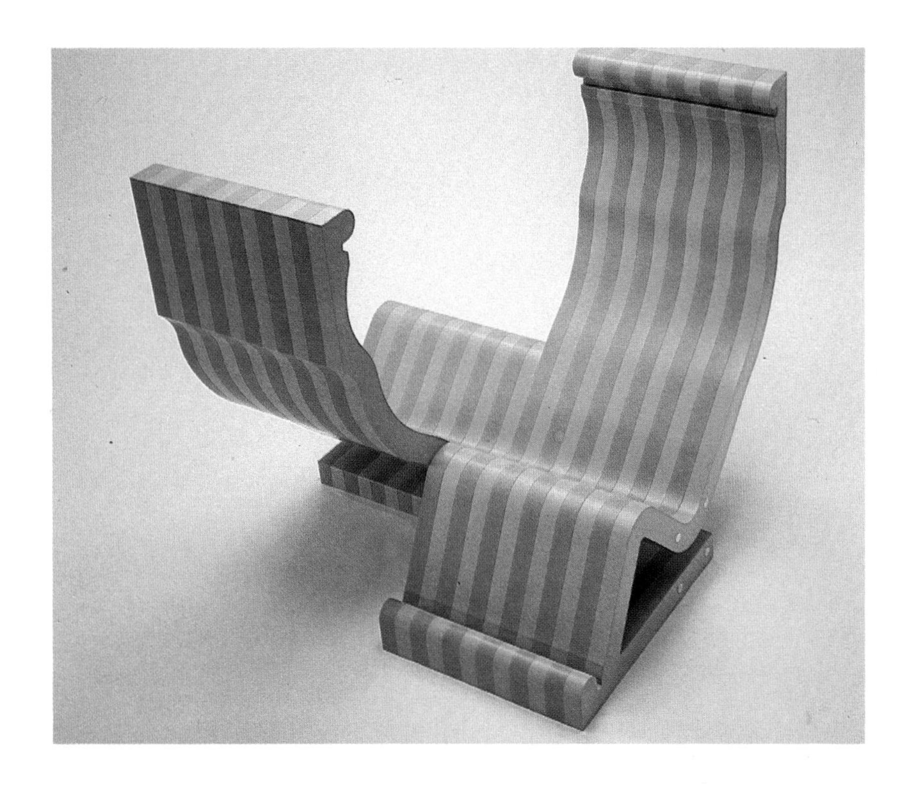

Product: Tête-à-Tête
Designer: Stanley Tigerman, Chicago, Illinois
Design Firm: Tigerman, Fugman, McCurry
Chicago Illinois
Designed for: Colorcore™ "Surface & Ornament" Competition I and Exhibition
Materials: Colorcore™ surfacing material by Formica Corporation

Product: L System
Designer: Emilio Ambasz and Giancarlo Piretti
New York, New York
Designed for: Colorcore™ "Surface & Ornament" Competition I and Exhibition
Materials: Colorcore™ surfacing material by Formica Corporation

Product: Modern Post-Neo Table
Designer: Milton Glaser, New York New York
Design Firm: Milton Glaser Associates, New York, New York
Designed for: Colorcore™ "Surface & Ornament" Competition I and Exhibition
Materials: Colorcore™ surfacing material by Formica Corporation

Product: Broken Length
Designer: Lella and Massimo Vignelli, New York, New York
Design Firm: Vignelli Associates, New York, New York
Designed for: Colorcore™ "Surface & Ornament" Competition I and Exhibition
Materials: Colorcore™ surfacing material by Formica Corporation

Product: Cart-Mobile
Designer: Ward Bennett, New York, New York
Designed for: Colorcore™ "Surface & Ornament" Competition I and Exhibition
Materials: Colorcore™ surfacing material by Formica Corporation

Product: Door
Designers: James Wines, Alison Sky
Michelle Stone, John deVitry,
New York, New York
Design Firm: SITE, New York, New York
Designed for: Colorcore™ "Surface & Ornament"
Competition I and Exhibition
Materials: Colorcore™ surfacing material by
Formica Corporation

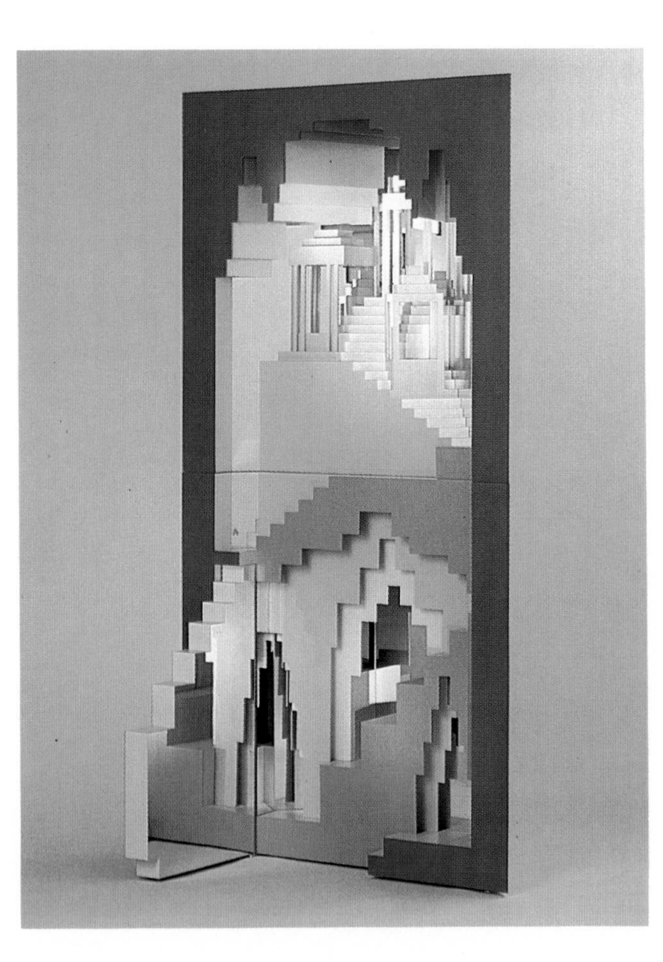

Product: Corner Cupboard
Designer: Charles Moore, Los Angeles, California
Design Firm: Moore, Ruble, Yudell, Los Angeles,
California
Designed for: Colorcore™ "Surface & Ornament"
Competition I and Exhibition
Materials: Colorcore™ surfacing material by
Formica Corporation

Product: Ryba
Designer: Frank Gehry, Los Angeles, California
Design Firm: Frank O. Gehry and Associates
Los Angeles, California
Designed for: Colorcore™ "Surface & Ornament" Competition I and Exhibition
Materials: Colorcore™ surfacing material by Formica Corporation

Product: Elevator Cab
Designer: Helmut Jahn, Chicago, Illinois
Design Firm: Murphy/Jahn, Chicago, Illinois
Designed for: Colorcore™ "Surface & Ornament" Competition I and Exhibition
Materials: Colorcore™ surfacing material by Formica Corporation

Product: Mirror in the Greek Revival manner
Design Firm: Venturi, Rauch, and Scott Brown
Philadelphia, Pennsylvania
Designed for: Colorcore™ "Surface & Ornament" Competition I and Exhibition
Materials: Colorcore™ surfacing material by Formica Corporation

Product: Latis Chair
Designer: James Evanson, New York, New York
Materials: Oak; lacquer; leather upholstery

Product: Michael Graves Seating Collection
Designer: Michael Graves, Princeton, New Jersey
Client: Sunar/Hauserman, Norwalk, Connecticut
Materials: Bird's-eye maple with upholstery

Product: Cinquecento
Designer: Alik Cavaliere
Client: Zanotta SPA, Milan, Italy
Materials: Surface: glass; Legs: burnished steel

Product: Imperiale
Designer: Achille Castiglioni
Client: Zanotta SPA, Milan, Italy
Materials: Frame: stainless steel; Armrests: wood; Seat: cotton fabric; upholstered headrest

Product: Cabaret
Designer: Stefano Casciani
Client: Zanotta SPA, Milan, Italy
Materials: Steel; polyurethane foam

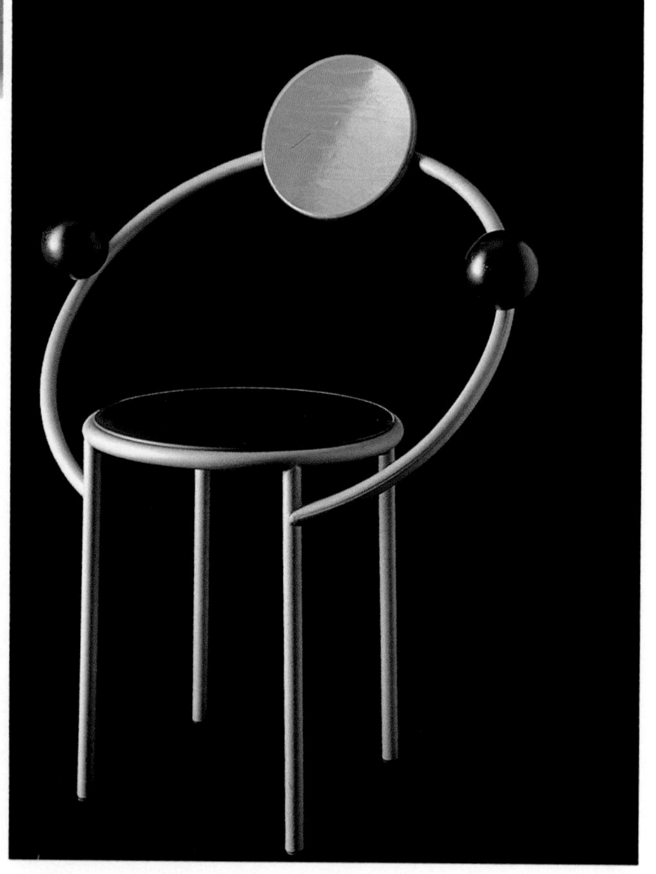

Product: First
Designer: Michele de Lucchi, Milan, Italy
Client: Memphis, Milan, Italy
Materials: Metal and wood

Product: Royal
Designer: Nathalie du Pasquier, Paris, France
Client: Memphis, Milan, Italy
Materials: Plastic laminate and printed cotton

Product: Experiment Collection
Designer: Yrjo Kukkapuro, Helsinki, Finland
Client: Avarte Oy, Helsinki, Finland
Materials: Seat and back: press-formed birch veneer, plastic laminate; Armrests: press-formed birch plywood; Legs: chrome-plated steel tube

Product: Sindbad
Designer: Vico Magistretti
Client: Cassina, Milan, Italy
Importer/Distributor: Atelier International, Ltd., New York, New York
Materials: Base: black lacquered beechwood; Frame: welded steel; Body: molded polyurethane foam and Dacron

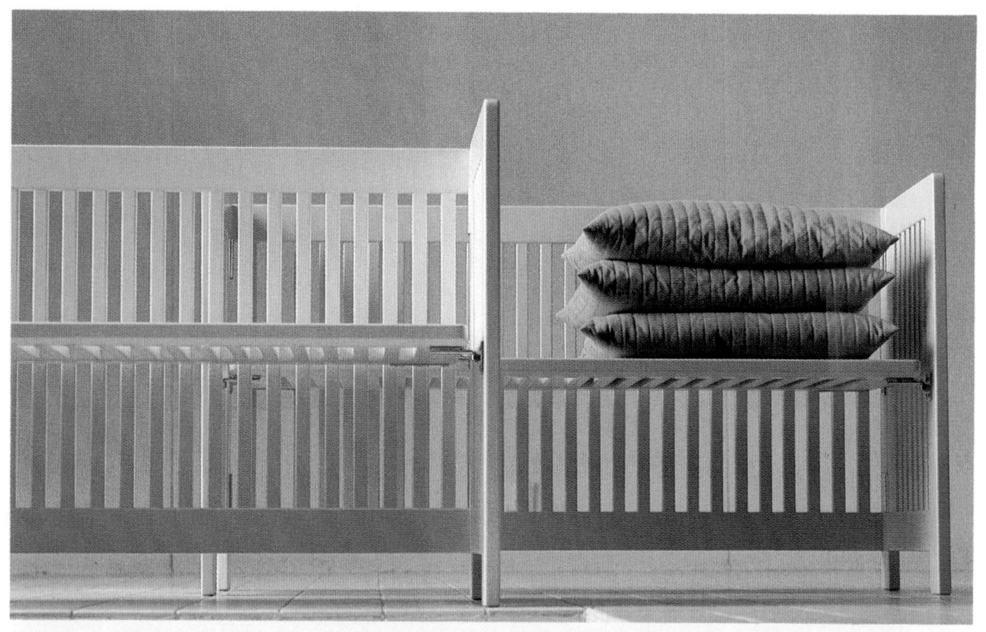

Product:	Painted wood garden seat
Designer:	Antti Nurmesniemi, Helsinki, Finland
Design Firm:	Studio Nurmesniemi, Helsinki, Finland
Client:	Vuokko Oy, Helsinki, Finland
Materials:	Red beech, painted white, brass fixtures

Product:	Steel wire garden seat
Designer:	Antti Nurmesniemi, Helsinki, Finland
Design Firm:	Studio Nurmesniemi, Helsinki, Finland
Client:	Vuokko Oy, Helsinki, Finland
Materials:	Epoxy-coated steel wire

Product:	Trapezoid
Designer:	James Evanson, New York, New York
Materials:	Plastic laminate with lacquer

Product:	Richard Meier Collection
Designer:	Richard Meier, New York, New York
Client:	Knoll International, New York, New York
Materials:	Maple with black, white, or natural hand-rubbed finish

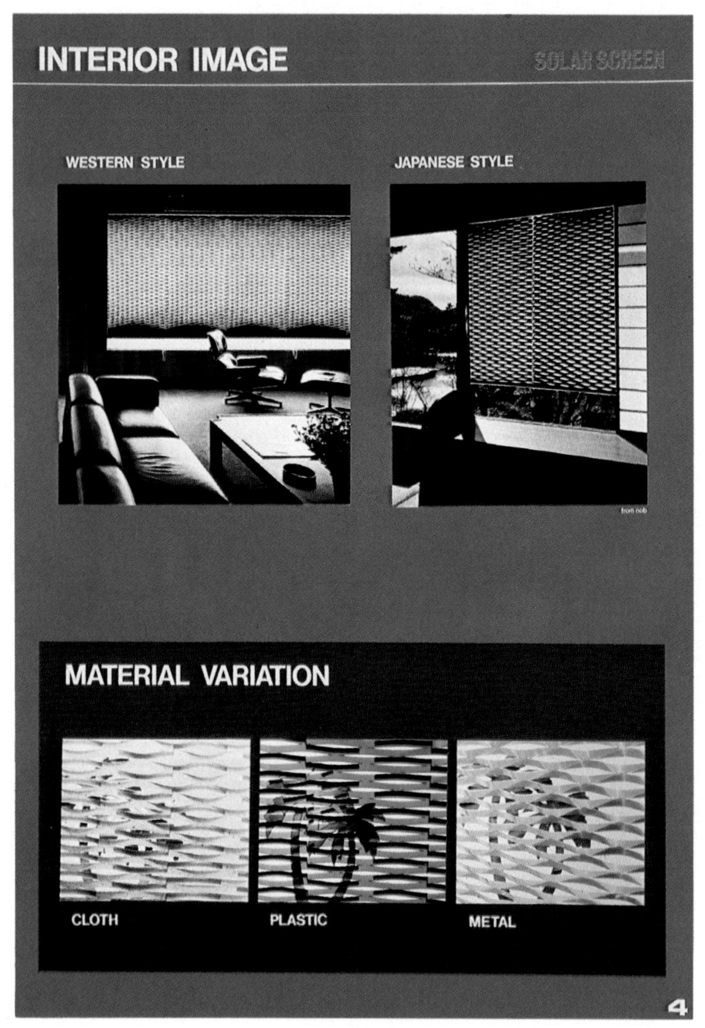

Product: Solar Screen
Designer: Tadahide Okuno, Osaka, Japan
Awards: 1st International Design Competition, Osaka

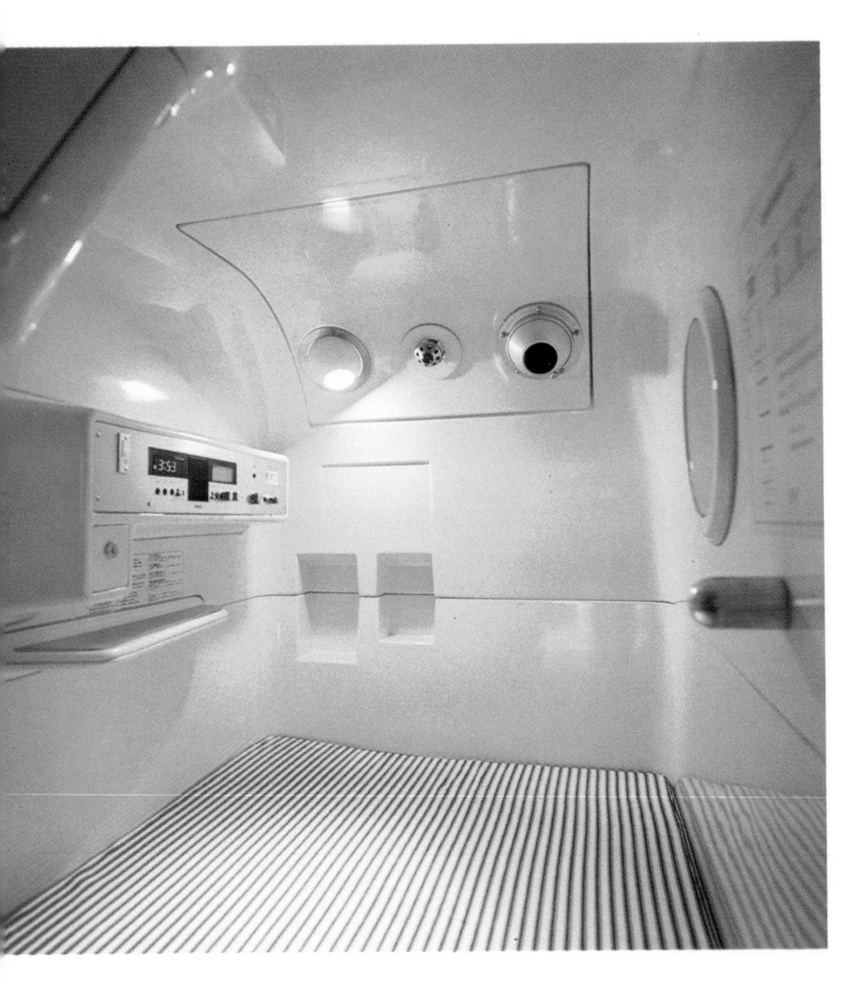

Product:	The Capsule hotel bed
Designer:	Kisho Kurokawa, Tokyo, Japan
Design Firm:	Kisho Kurokawa Architect & Associates, Tokyo, Japan
Client:	Kotobuki Seating Co. Ltd., Tokyo, Japan
Materials:	Phenol resin modules molded in two halves and bolted together and stacked on steel frames

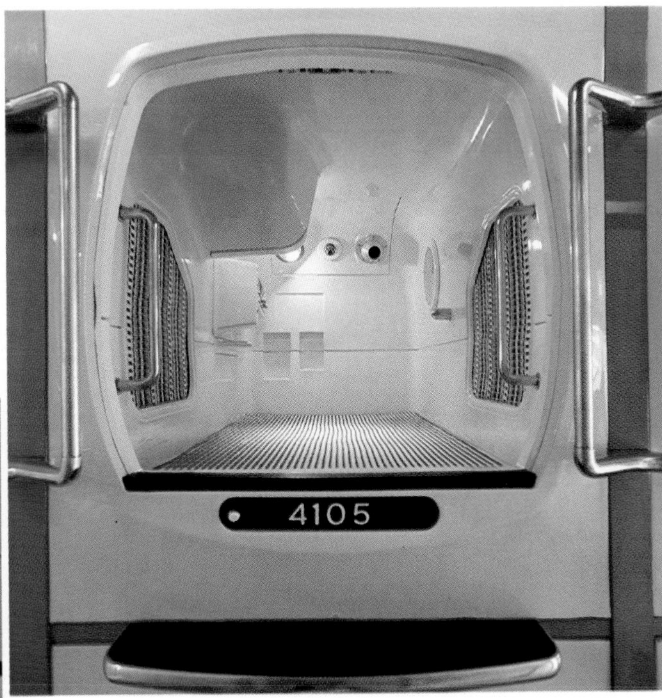

CHAPTER **5**

CESHOUSEWARESTOOLSAPPLIANCESHOUSEWARESTOOLSAPPLIANCESHOUSEWARESTOOLSAPPLIANCESHOUSEWAREST
CESHOUSEWARESTOOLSAPPLIANCESHOUSEWARESTOOLSAPPLIANCESHOUSEWAR

CTRONICSENTERTAINMENTHOMEELECTRONICSENTERTAINMENTHOMEELECTRON
INMENTHOMEELECTRONICSENTERTAINMENTHOMEELECTRONICSENTERTAINMEN

GLIGHTINGLIGHTINGLIGHTINGLIGHTINGLIGHTINGLIGHTINGLIGHTINGLIGHTINGLIGHTINGLIGHTINGLIGHTING
GLIGHTINGLIGHTINGLIGHTINGLIGHTINGLIGHTINGLIGHTINGLIGHTING

TRESIDENTIALFURNISHINGSCONTRACTRESIDENTIALFURNISHINGSCONTR
TIALFURNISHINGSCONTRACTRESIDENTIALFURNISHINGSCONTRACTRESID

Business Equipment

SEQUIPMENTBUSINESSEQUIPMENTBUSINESSEQUIPMENTBUSINESSEQUIPMENTBUSINESSEQUIPMENTBUSINESSEQU
SEQUIPMENTBUSINESSEQUIPMENTBUSINESSEQUIPMENTBUSINESSEQUIPMENTBUSINESSEQUIPMENTBUSINESSEQU

EQUIPMENTMEDICALEQUIPMENTMEDICALEQUIPMENTMEDICALEQUIPMENTMEDICALEQUIPMENTMEDICALEQUIPMENT
NTMEDICALEQUIPMENTMEDICALEQUIPMENTMEDICALEQUIPMENTMEDICALEQUIPMENTMEDICALEQUIPMENTMEDICAL

IALEQUIPMENTTRANSPORTATIONINDUSTRIALEQUIPMENTTRANSPORTATIONINDUSTRIALEQUIPMENTTRANSPORTAT
IALEQUIPMENTTRANSPORTATIONINDUSTRIALEQUIPMENTTRANSPORTATIONINDUSTRIALEQUIPMENTTRANSPORTAT

IONALSPORTSEQUIPMENTRECREATIONALSPORTSEQUIPMENTRECREATIONALSPORTSEQUIPMENTRECREATIONALSPO
NTRECREATIONALSPORTSEQUIPMENTRECREATIONALSPORTSEQUIPMENTRECREATIONALSPORTSEQUIPMENTRECREA

STEXTILESTEXTILESTEXTILESTEXTILESTEXTILESTEXTILESTEXTILESTEXTILESTEXTILESTEXTILESTEXTILES
STEXTILESTEXTILESTEXTILESTEXTILESTEXTILESTEXTILESTEXTILESTEXTILESTEXTILESTEXTILESTEXTILES

FORTHEHANDICAPPEDDESIGNSFORTHEHANDICAPPEDDESIGNSFORTHEHANDICAPPEDDESIGNSFORTHEHANDICAPPED
ANDICAPPEDDESIGNSFORTHEHANDICAPPEDDESIGNSFORTHEHANDICAPPEDDESIGNSFORTHEHANDICAPPEDDESIGNS

Perhaps no category more than that of business equipment exemplifies how the design industry has benefited from the recent and sudden growth of communications technology. As telecommunications have made the world a smaller place, it is designers who have helped to make this transition a comfortable one. As the complexities and capabilities of business equipment increase, so too, in direct proportion, must their simplicity and ease of function. More and more, it is a case of simplicity undercutting complexity.

Ease of operation is vital not only because of the growing capabilities of much of this equipment, but also because of the varied levels of electronic literacy of its users. Witness, for example, the Edge videotape editing console designed by Steinhilber & Deutsch, Associates. Part of the console's success is due to that fact that it is easily operated by users with any level of experience. Because its design helps to make its operation self-evident, the inexperienced user is able to concentrate on editing, rather than on the mastery of the controls.

Aside from this rather obvious point of ease of use, another apparent feature of contemporary business equipment is its compactness. Designer input is integral both to equipment which combines different communications functions and equipment which has simply become smaller, or even miniaturized. The growth in the variety of products which can be accommodated by the briefcase is matched in number only by the challenges they present to the designer.

A second, and no less important feature is compatibility. Whether it is the business itself that is expanding, or the business's need and use of electronics, or both, component systems which are designed to be updated to accommodate new equipment have a clear edge. And, as electronic literacy is on the rise, users demand new capabilities. Systems which can respond to this sophistication are meeting a real market demand. Compactness and compatibility—often approached simultaneously—are surely the two keynotes in the design of contemporary business equipment, and computer design is, of course, the most active arena for their development.

The MAD-1 microcomputer (the acronym is for Modular Advanced Design) designed by Mike Nuttall is a case in point. The unit is among the first small computers to use the powerful 80186 microprocessor—which has twice as many transistors in its quarter-inch silicon chip as its predecessor, the 8088. The 80186 integrates the fifteen to twenty chips previously required into a single integrated circuitry, in effect installing a full circuit board into a miniature case. But as significant is the fact that the 80186 can be used with the same instructions as the 8088. So, users of the MAD-1 can use most of the same software, in particular the business applications software designed for the IBM PC, but twice as fast. In addition, the compact MAD-1 is ergonomically designed: The slope of the low profile keyboard is reduced, the display monitor tilts and swivels to reduce eye and neck fatigue, a palm rest area supports the hands, and the modularity of the systems permits different configurations on the desk top.

The designers at I D Two of Grid's Compass portable computer faced even more extreme challenges. As the firm's director, Bill Moggridege, explains, "New technologies made a truly portable computer possible. Flat panel displays, bubble memory, low-profile keyboards, large-scale integration of logic, and small modems for the transmission of both voice and data via telephone lines all made it possible to develop for the first time a powerful machine which could fit into a briefcase." A team comprising both systems engineers and industrial designers developed a unit that weighed 9½ lbs. (4.3 kg.) and measured 11½" × 15½" × 2" (2 × 39 × 5 cm.). Its computing capabilities, meanwhile, were powerful enough to handle budgetary planning and, through the telephone line, could also retrieve programs, access centrally stored data, and send and receive messages by electronic mail. The housing for the entire unit was magnesium, a deviation from the standard plastic housing of portable computers, that was selected for its light weight and adaptability to precision tooling—which permitted additional design flexibility.

What distinguishes the Compass as much as its unconventional housing, advanced computing power in a portable size, and refined styling is the manner in which it was developed. Participating in its development were an industrial designer, a mechanical engineer, an electronics systems engineer, a logic and wiring engineer, and a staff member conversant with manufacturing processes. Because the unit does indeed "push packaging technology," the engineers recognized the need for industrial and mechanical design staff to collaborate with electronics logic experts. That industrial designers should work so closely with engineers in the start-up phase of product development—to determine performance as well as appearance—is indeed unique.

While larger business equipment manufacturers, such as IBM, have traditionally been conspicuous for the high-quality design of their products, smaller firms have rarely followed suit. The electronics industry—marked by the recent and rapid emergence of small firms—however, has been turning all this around by noting the fact that competitive design work in product development may be what determines the leading edge on the market. In judging the many innovations occurring in California's Silicon Valley, this re-evaluation of the designer/engineer relationship is not the least insignificant.

Clearly, the domain of the industrial business equipment designer has grown in recent years. Color, texture, material, and system image are no longer the only areas that answer to his specifications. His considerations have grown to include size, ergonomic adjustability, modularity within the system, and compatibility to other systems, to name a few. And as design in this category becomes more an issue of product development than of styling, it also becomes a research-based process. But on the other end, these same designers are becoming involved with marketing their products. Because the design strategy begins at an earlier stage, it is not surprising that it should be continued into a later one as well. That is, as the design strategy becomes more integral to the overall business strategy of a product, it has as much a place in marketing as it does in product development. Out of necessity, designers working in this field of equipment have become literate in new areas, recognizing the fact that sales may come to depend upon their recognition of design as an interdisciplinary medium. What all of this suggests is not simply that the context of the design profession has grown here, but that its lead may well be followed elsewhere.

Product: PCS-Cadmus 2200 terminal
Designer: Alexander Neumeister
Munich, West Germany
Design Firm: Neumeister Design
Munich, West Germany
Client: Periphere Computer Systeme GmbH
Munich, West Germany
Materials: Anti-glare display unit; low contrast casing; tilting mechanism; laser beam printer

Product: Mad-1 computer
Designer: Mike Nuttall
Palo Alto, California
Design Firms: ID Two
and Dave Kelley Design
Palo Alto, California
Client: Mad Computer Inc.
Santa Clara, California
Materials: ABS housing
Photo Credit: Jack Christianson

Product: Pronto Series 16 computer
Design Firm: Ron Loosen Associates
Los Alamitos, California
Client: Pronto Computers
Torrence, California
Awards: 1983 *Industrial Design* magazine
Design Review selection
Materials: Sheet metal with foam front electronics box; entire unit is painted in polane

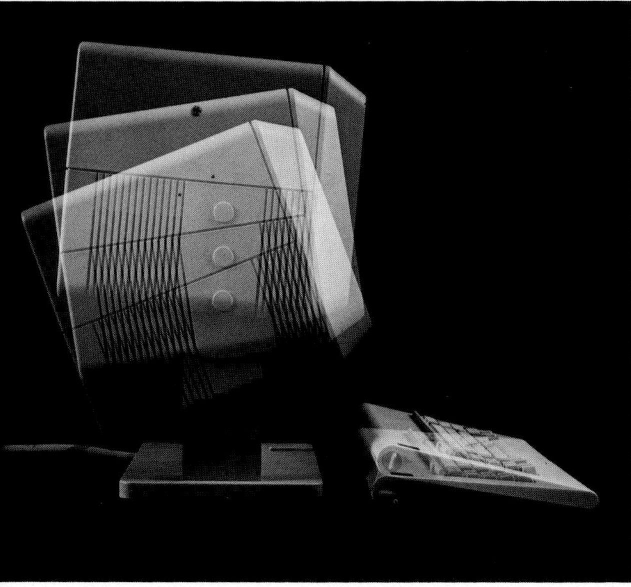

Product: IBM System 23
Designer: Tom Hardy
Lexington, Kentucky
Client: IBM Corporation
Lexington, Kentucky
Awards: 1983 *Industrial Design* magazine
Design Review selection
Materials: Sheetmetal; structural foam; ABS plastic

Product: Phaze P3278 computer terminal
Design Firm: Fischer-Design
Scottsdale, Arizona
Client: Phaze Information Machines
Scottsdale, Arizona
Awards: 1983 *Industrial Design* magazine
Design Review selection
Materials: Monitor and keyboard: injection-molded
Noryl and die-cast aluminum

Product: Compass computer
Design Firm: ID Two
Palo Alto, California
Client: Grid Systems
Mountain View, California
Awards: 1982 IDSA Industrial Design Excellence Award
Materials: Die-cast magnesium and injection-molded plastic

Product: StacPac Modules
Designer: Mike Nuttall
Palo Alto, California
Design Firm: I D Two
Palo Alto, California
Client: Data Systems Design
San Jose, California
Awards: 1983 *Industrial Design* magazine Design Review selection
1983 SMAU Industrial Design Award
Materials: Front and rear bezels: injection-molded texture plastic; aluminum body is painted with a powder coating to match front and rear bezels

Product: WY-1000 terminal
Designer: Jim Sacherman
Palo Alto, California
Design Firm: Matrix Product Design, Inc.
Palo Alto, California
Client: Wyse Technology
San Jose, California
Materials: Low-profile sculptures keyboard
Photo credit: Henrik Kam

Product: ITT 3840 computer work station and disc module
Design Firm: Fischer-Design
Scottsdale, Arizona
Client: Standard Elektrik Lorenz, subsidiary of ITT
Munich, West Germany
Materials: Fire-retardant Noryl structural foam

Product: M 10 portable computer
Designers: Perry A. King and Antonio Macchi Cassia
Milan, Italy
Client: Ing. C. Olivetti & C. S.p.A.
Ivrea, Italy
Awards: 1983 SMAU Design Award

Product: VME/10 microcomputer system
Design Firm: Fischer-Design
Scottsdale, Arizona
Client: Motorola Micro Systems
Phoenix, Arizona
Materials: Fire-retardant Noryl structural foam; sheetmetal

Product: CIE-7800 multi-purpose computer terminal
Design Firm: Fischer-Design
Scottsdale, Arizona
Client: C ITOH
Tokyo, Japan
Materials: High-pressure injection-molded ABS

Product: Compact Computer 40
Designer: Consumer Design Center,
Texas Instruments Inc.
Lubbock, Texas
Client: Texas Instruments Inc.
Lubbock, Texas
Materials: Top case: molded ABS plastic; keys: double-shot injection molded ABS; base; anodized aluminum

Product: Wang Professional Computer
Designers: Karen L. Usab, Lawrence M. Kuba, Douglas C. Dayton, and Paul W. Porter
Lowell, Massachusetts
Client: Wang Laboratories Inc.
Lowell, Massachusetts
Materials: Sheet metal CPU housing; structural foam molded monitor base; cast aluminum monitor arm; extruded aluminum desk clamp; monitor and keyboard housing and CPU front bezel: injection-molded

Product: WorkSlate
Designer: Mike Nuttall
Palo Alto, California
Design Firm: Matrix Product Design Inc.
Palo Alto, California
Client: Convergent Technologies, Inc.
Santa Clara, California
Materials: Injection-molded ABS
Photo Credit: Light Language, San Francisco

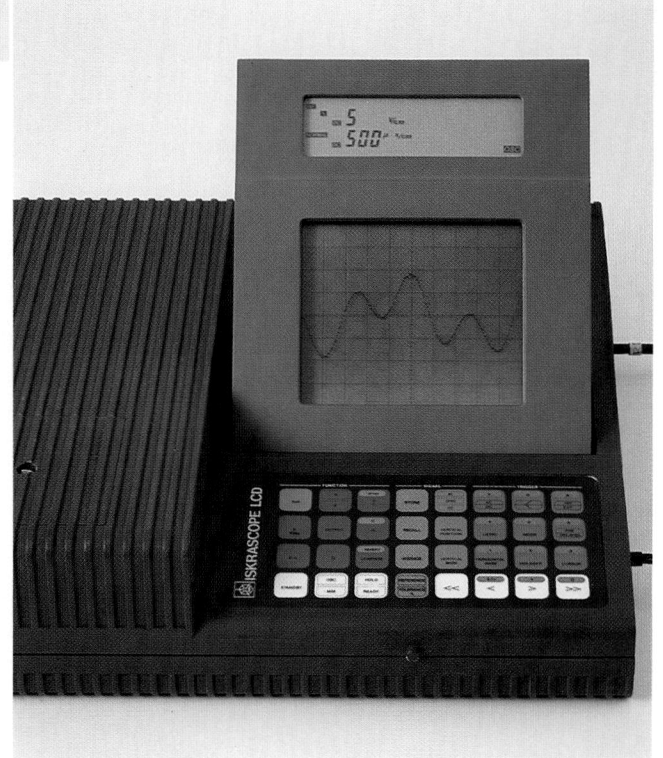

Product: Iskrascope LCD—digital osciloscope, digital multimeter, data processor, and transient recorder
Designer: Ljuban Klojcnik
Kranj, Yugoslavia
Design Firm: Iskra Kibernetika
Kranj, Yugoslavia
Client: Iskra Commerce
Ljubljana, Yugoslavia
Materials: ABS plastic

Product: Modular school computer system prototype
Designer: Volker Zolch
Essen, West Germany
Awards: 1983 Braun Prize for Technical Design

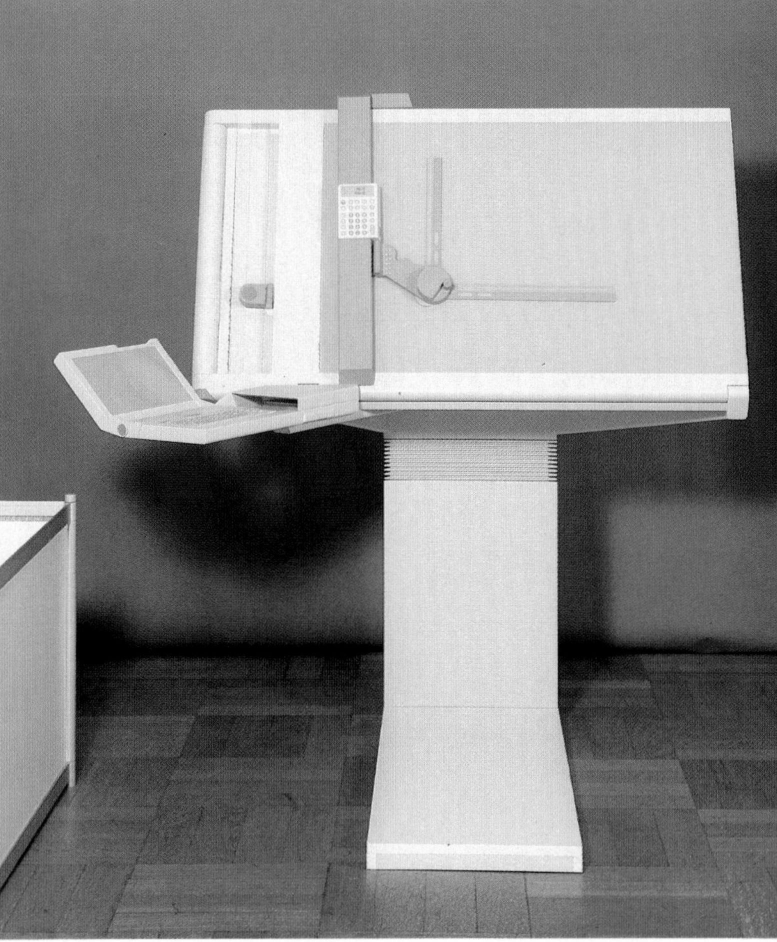

Product: CAD System prototype
Designer: Uwe Kemker
Iserlohn, West Germany
Awards: 1983 Braun Prize for Technical Design

Product: NCR 2126 retail checkout system
Designer: Sangyo Design, Zip Co., Ltd.
Kanagawa, Japan
with Corporate Industrial Design, NCR Corporation
Dayton, Ohio
Client: NCR Corporation
Dayton, Ohio

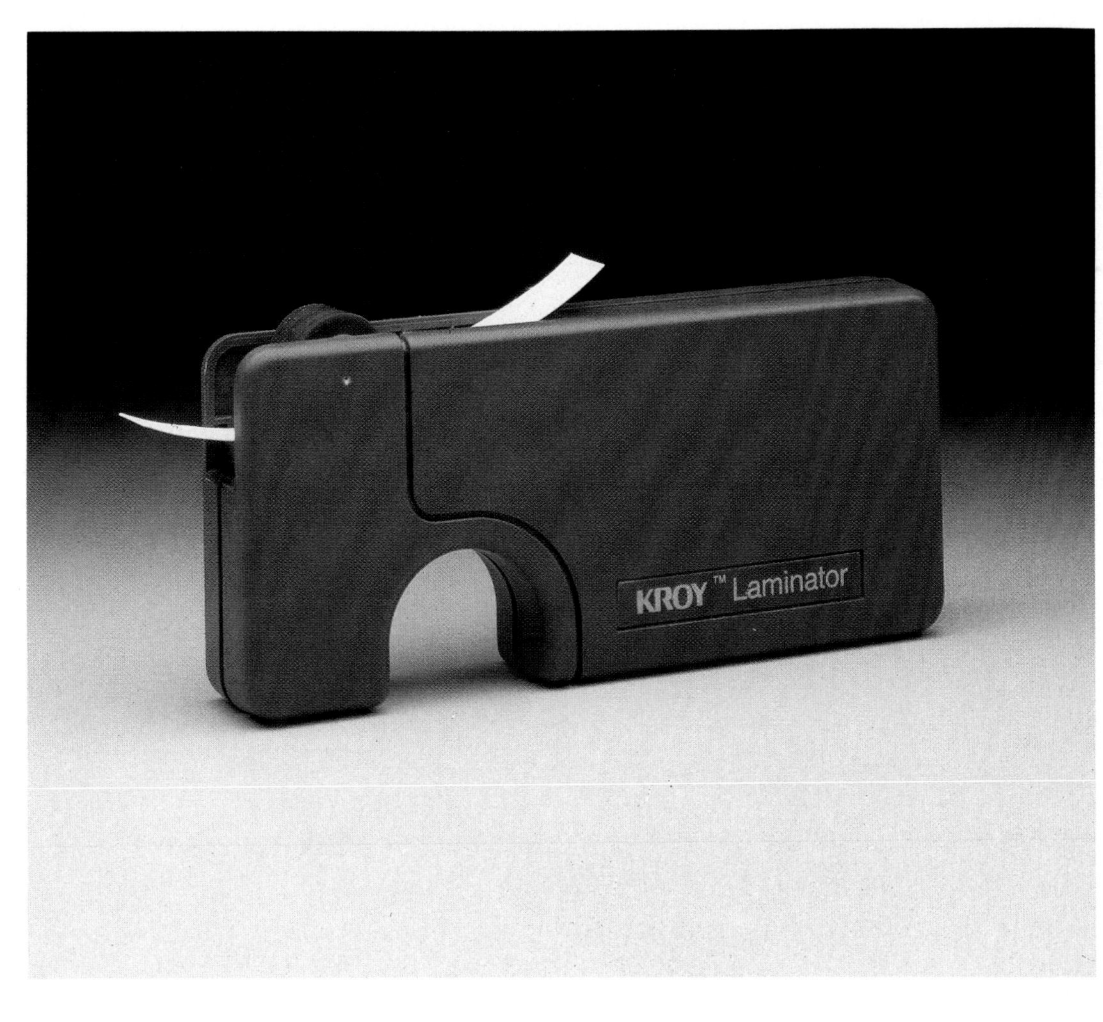

Product: Kroy laminator
Designer: W. Robert Worrell
Minneapolis, Minnesota
Design Firm: Worrell Design Incorporated
Minneapolis, Minnesota
Client: Kroy Inc.
St. Paul, Minnesota
Awards: 1982 IDSA Industrial Design Excellence Award
Materials: Injection-molded, flame retardent ABS; sonically welded

Product: LetraGraphix typesetter
Designers: Davin Stowell Associates
New York, New York
with Engineered Plastic Products
Stirling, New Jersey
Client: Esselte Letraset Manufacturing
Moonachie, New Jersey
Awards: 1983 *Industrial Design* magazine Design Review selection
Materials: Plastic grooved magnet mounted on a four-piece thermoformed polystyrene housing; hinges: polypropylene; ruler: acrylic; components assembled with pressure-sensitive adhesive

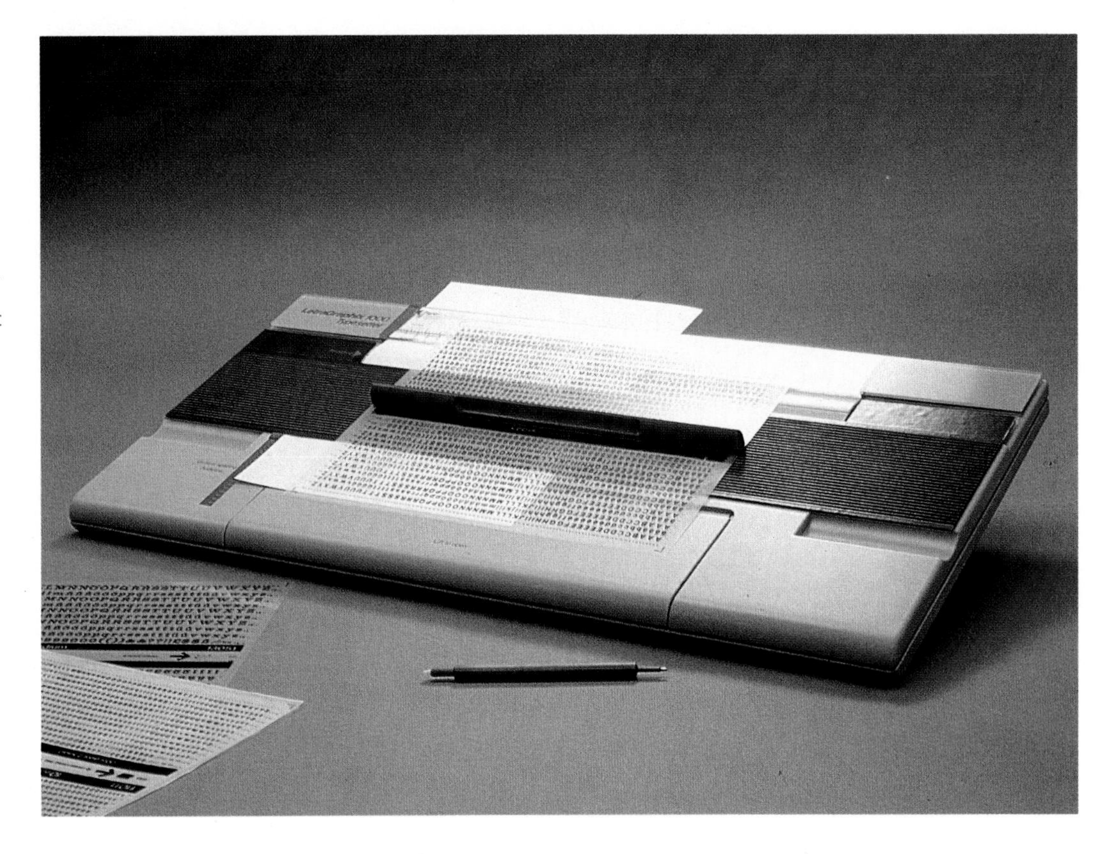

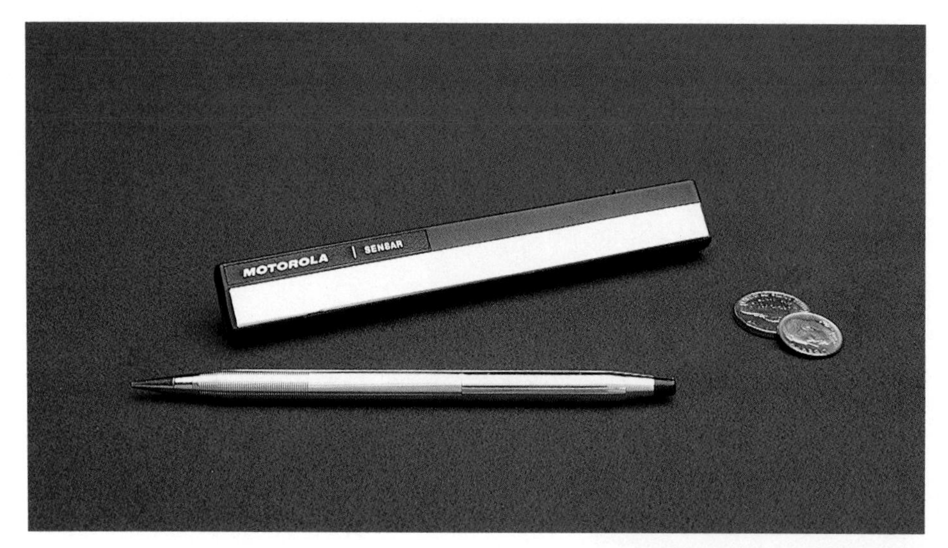

Product: Sensar FM radio pager
Designers: Paging Division, Motorola Inc.
Fort Lauderdale, Florida
Client: Motorola Communications and Electronics, Inc.
Schaumburg, Illinois
Materials: Housing: polycarbonate; antenna: phosphorous bronze, chrome-plated strip; clip: aluminum

Product: Point-of-Scale terminal
Design Firm: Fischer-Design
Scottsdale, Arizona
Client: National Semiconductor Datachecker/DTS
Sunnyvale, California
Materials: Painted aluminum die castings and high-impact self-colored polycarbonate

Product: Norelco MicroSeries Microcassette Recorder/Transcriber MCR-7200
Design Firm: Michael W. Young Associates Incorporated
Flushing, New York
Client: Philips Business Systems, Inc.
Woodbury, New York
Materials: Injection-molded ABS; semi-matte finish with glossy finish detailing

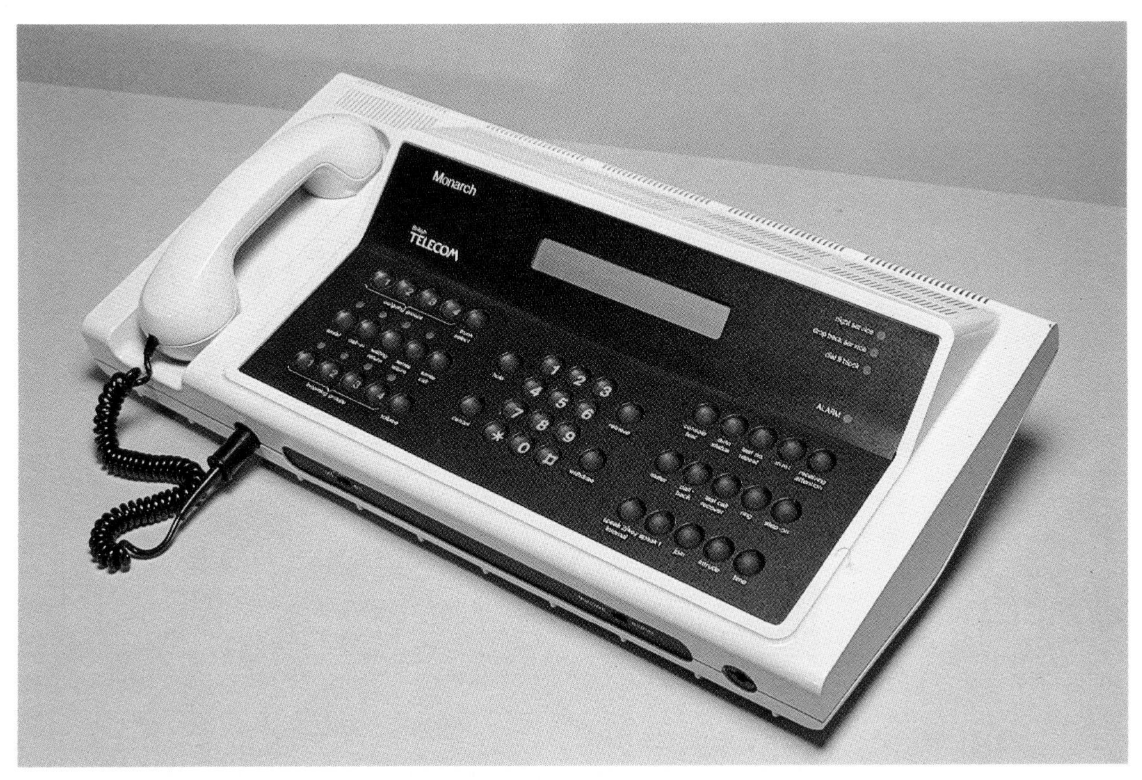

Product: Monarch 120 Call-Connect system
Design Firm: DCA Design Consultants
London, England
Client: British Telecom
London, England
Awards: 1983 Design Council Award
Materials: Black keyboard with touch-sensitive depressions; ergonomically designed, without plugs, switches, or buttons; power supply and control equipment fit into a single cabinet

Product: Teletracer 2800 pager
Design Firm: Industrieel Ontwerpburo Berkheij
Driebergen, The Netherlands
Client: Nira International
Emmen, The Netherlands
Awards: 1983 *Industrial Design* magazine Design Review selection
Materials: Receiver: liquid crystal display; microprocessor: C-MOS technology; keyboard with rubber switches

Product: Oce 1900 plain paper copier
Designer: Louis Lucker, Oce Design Team
Client: Oce Netherlands B.V.
Venlo, The Netherlands
Materials: Sheet steel; modified PPO foam

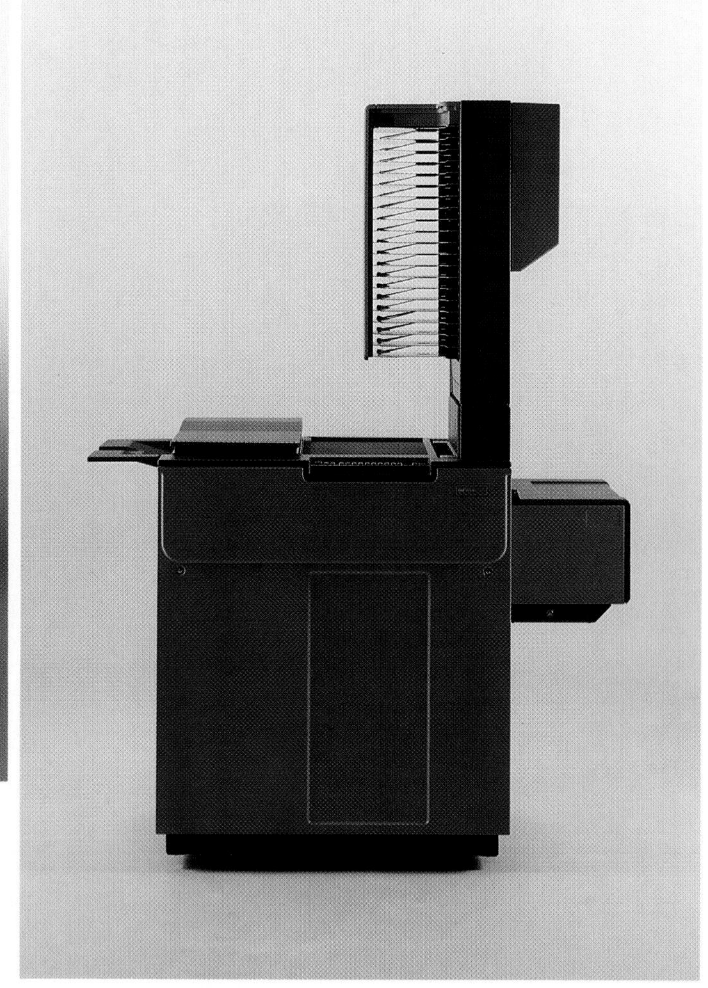

Product: Oce 1725 plain paper copier
Designer: Louis Lucker, Oce Design Team
Client: Oce Netherlands B.V.
Venlo, The Netherlands
Materials: Sheet steel; modified PPO foam

Product: Beta 450Z copier
Designers: Minolta Engineering Department
Osaka, Japan
Client: Minolta Corporation
Ramsey, New Jersey
Materials: Injection-molded ABS plastic

Product: Oce 3760 Electro Static A1 Micro Enlarger/Printer
Designer: Louis Lucker, Oce Design Team
Client: Oce Netherlands B.V.
Venlo, The Netherlands
Materials: Sheet steel

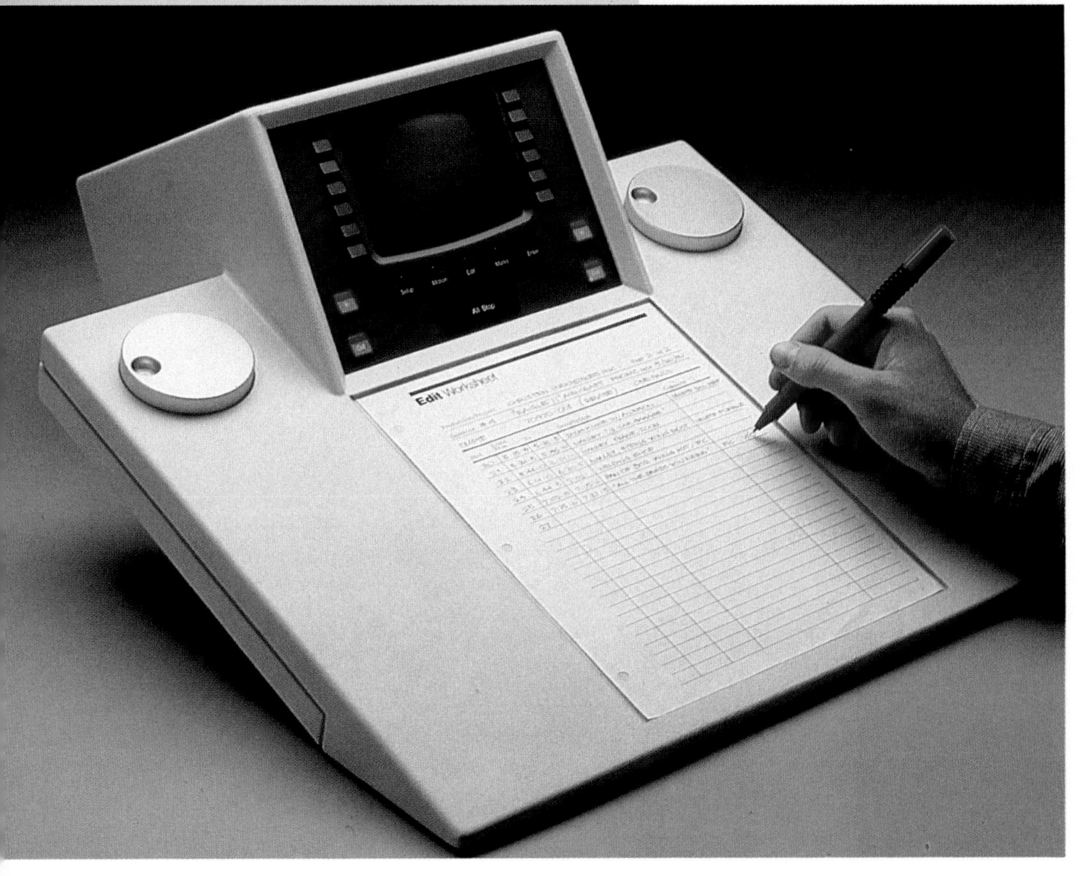

Product: CMX Systems videotape editor console
Design Firm: Steinhilber & Deutsch Associates
San Francisco, California
Client: CMX Systems, Orrox Corporation
Santa Clara, California
Awards: 1981 *Industrial Design* magazine
Design Review selection
Materials: Housing: structural polyurethane foam; tape position control wheels: aluminum; housing finish: two-part urethane enamel; bezel plate: urethane enamel with satin finish; control wheels: clear anodized finish

Product: Micronunciator alarm annunciator
Designers: D. Celine, M. Gans, L. Thole, and design team,
Control Logic (Pty) Ltd.
Client: Control Logic (Pty) Ltd.
Durban, South Africa
Awards: 1982 Shell Design Award

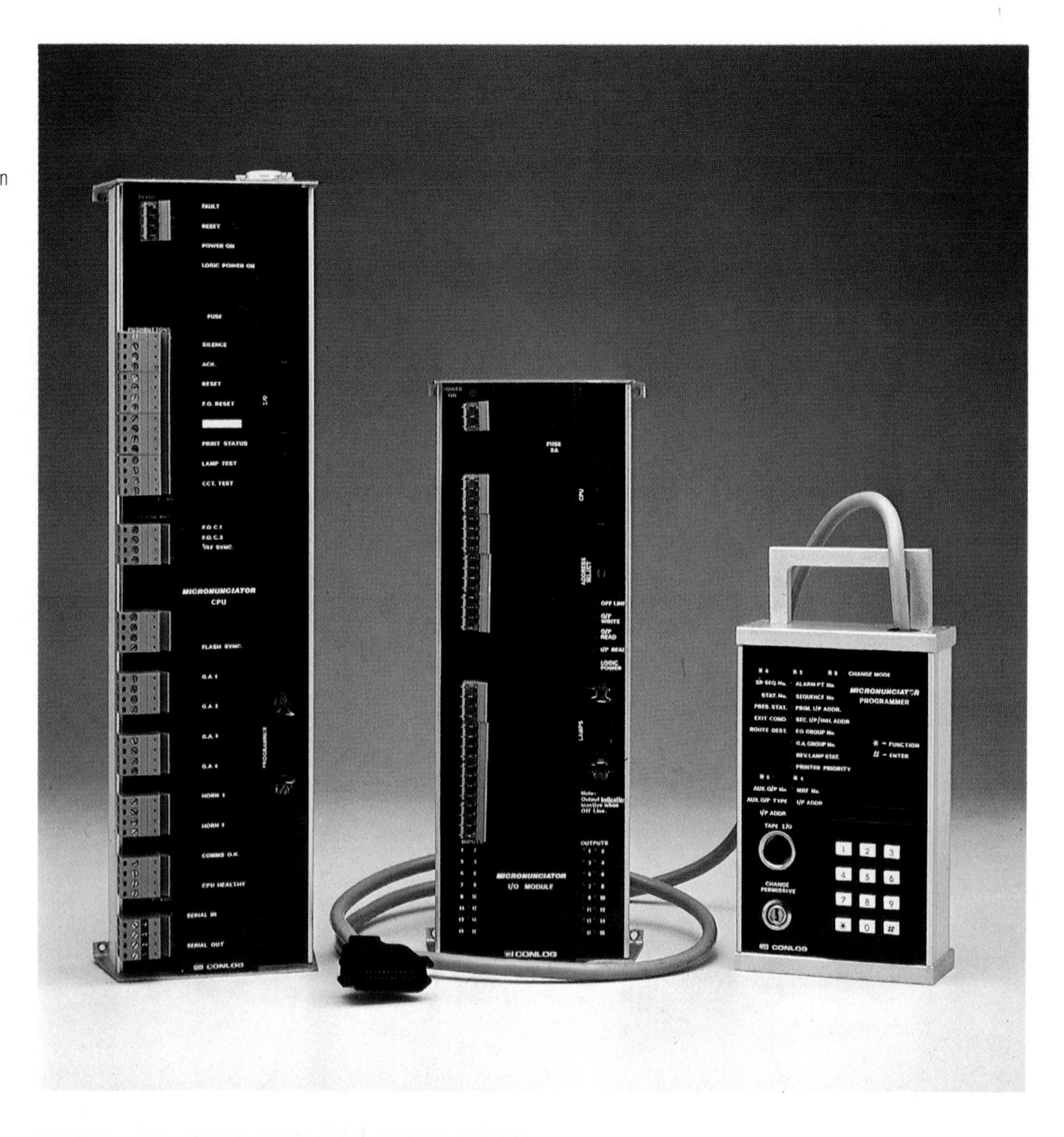

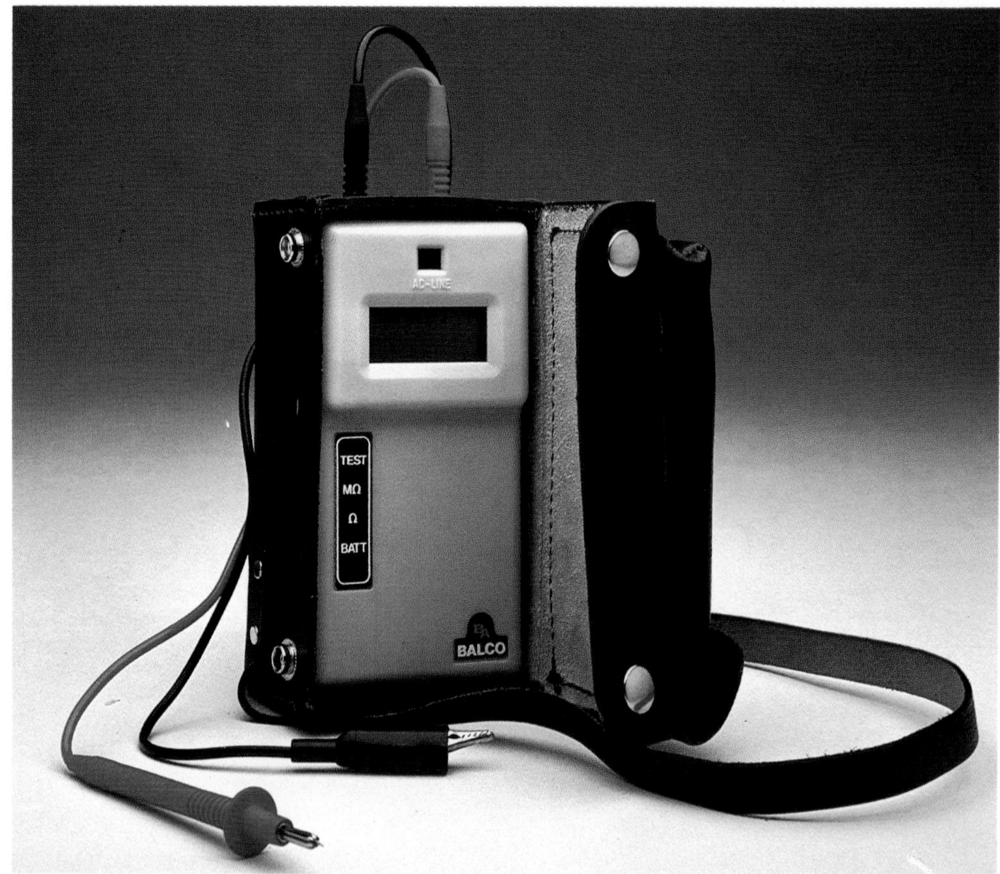

Product: Omeg/Balco digital insulation and continuity tester
(This unit is used for installation and maintainence of domestic
and industrial wiring layouts)
Designer: K. W. Junker and H. M. Teubner
Client: Omnitec Institute and Electronic Manufacturers (Pty) Ltd.
Excom, South Africa
Awards: 1982 Shell Design Award
Materials: Polycarbonate casing

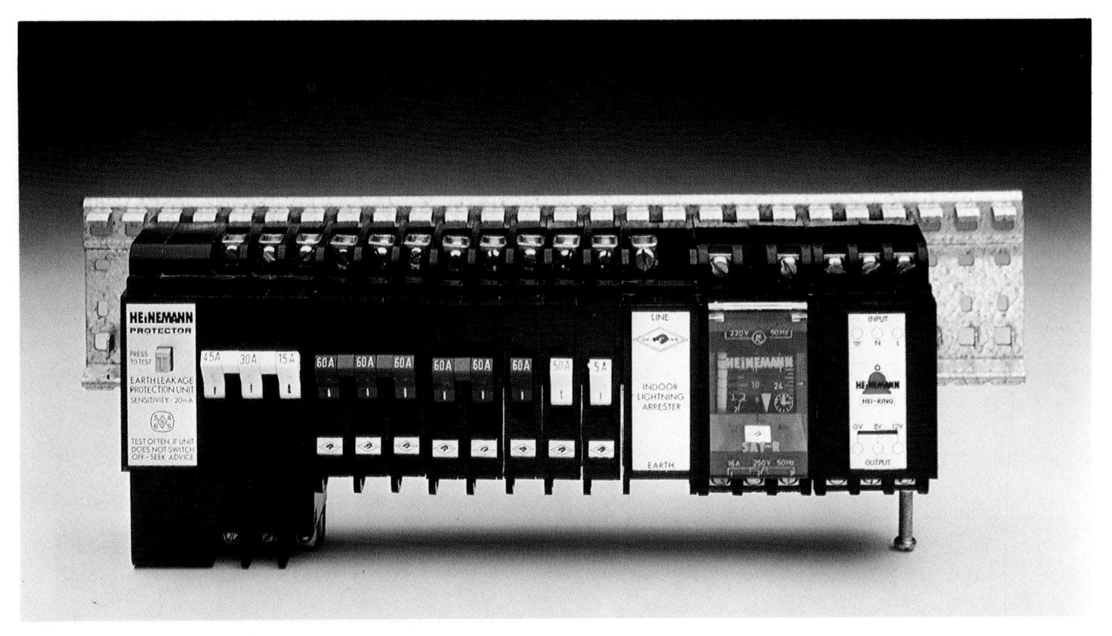

Product: Samite electrical protection device
Designers: K. Nüsse, G. Campetti, H. F. Andrews, N. Baumgarti, M. Ribeiro, A. S. Drummond, A. Tsiaparis, P. Baines, S. W. Clives, and design team
Client: Heinemann Electric (SA) Ltd. Johannesburg, South Africa
Awards: 1982 Shell Design Award

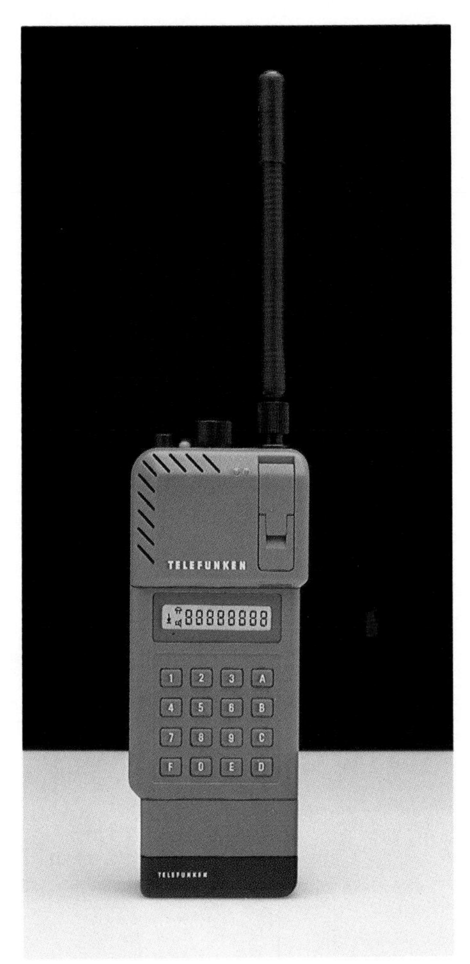

Product: Teleport 9 professional walkie-talkie
Design Firm: frogdesign Campbell, California
Client: AEG Telefunken West Germany
Materials: Aluminum castings

Product: Library Laser
Designer: William Sklaroff
Design Firm: William Sklaroff Design Associates Philadelphia, Pennsylvania
Client: Metrologic Instruments Inc. Bellmaur, New Jersey
Materials: Side panels: aluminum; base: Corian; housing: ABS Cycolac; finish: steel with baked enamel

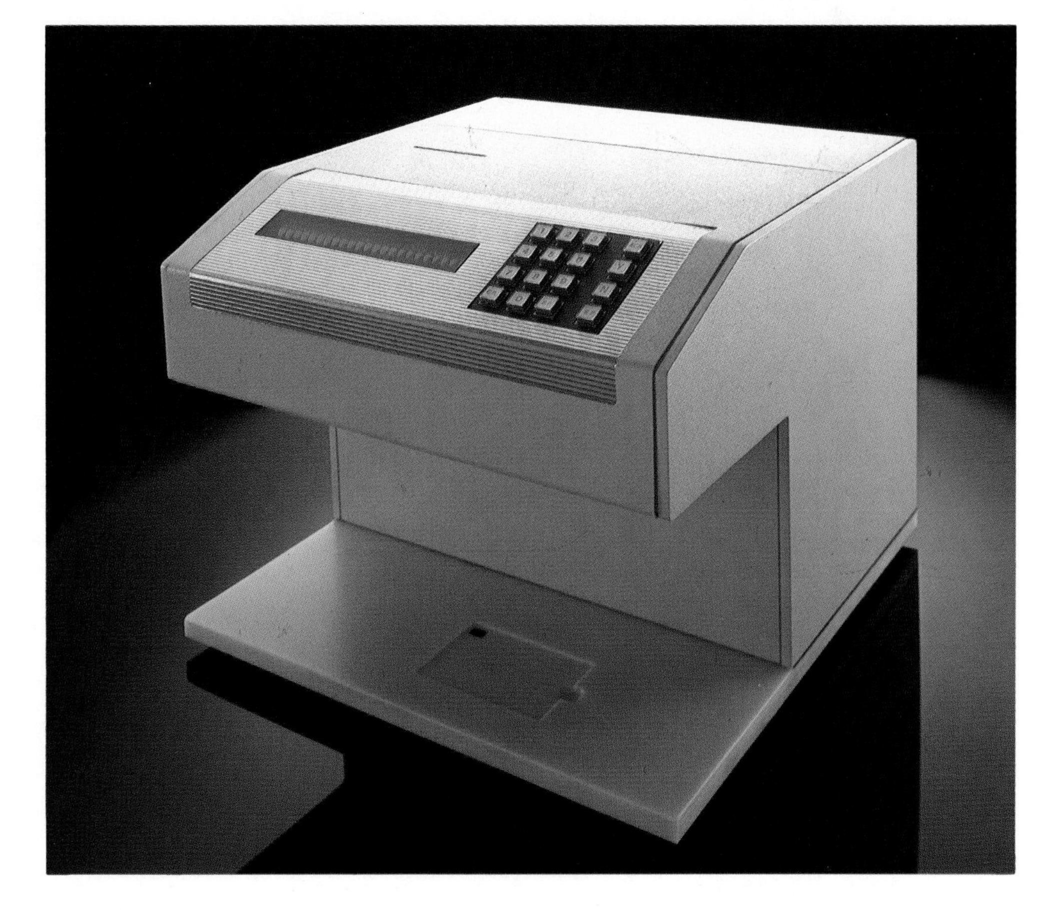

Product: Y-688^{32} total error corrector video processor
Designer: Bruce Pharr, design staff
Fortel Inc.
Design Firm: Penney & Bernstein
New York, New York
Client: Fortel Inc.
Atlanta, Georgia
Materials: Aluminum housing

Product: Disa Telephone 1200 Type Range
Designers: K. Hartiani, R. Derbyshire and design team,
Telephone Manufacturers of South Africa;
J. A. Raath and B. A. Bets, and design team,
Department of Posts and Telecommunications;
Roger Williams Associates
Pretoria, South Africa
Client: Telephone Manufacturers of South Africa
Springs, South Africa
Awards: 1982 Shell Design Award

Product: Handheld laser
Designer: William Sklaroff
Design Firm: William Sklaroff Design Associates
Philadelphia, Pennsylvania
Client: Metrologic Instruments Inc.
Bellmaur, New Jersey
Materials: ABS Cycolac housing

Product: Record 2 video detail booster
Designer: Bruce Pharr, design staff
Fortel Inc.
Design Firm: Penney & Bernstein
New York, New York
Client: Fortel Inc.
Atlanta, Georgia
Materials: Aluminum housing

Product: Cafe-Bar Services
Client: Burns, Philp & Company Limited
Sydney, Australia

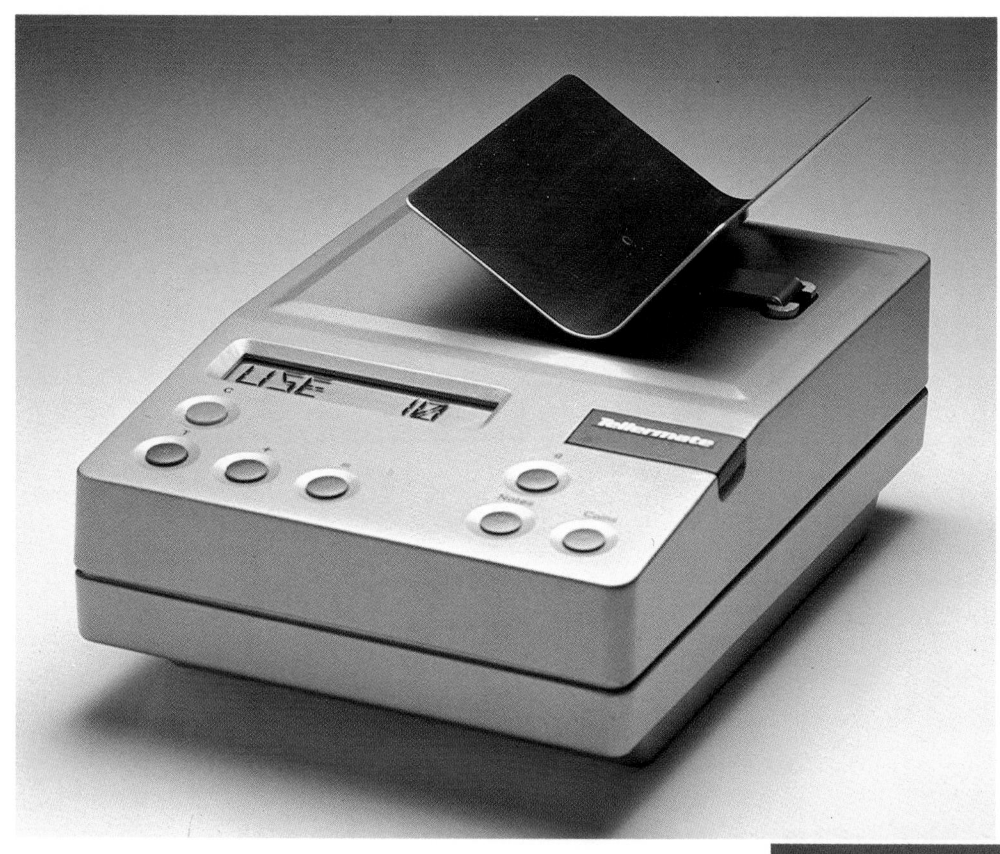

Product: Tellermate currency counter
Designer: Nicholas Oakley
London, England
Client: Perkam Ltd.
London, England
Awards: 1982 *Industrial Design* magazine
Design Review selection
Materials: Housing: structural foam; hood: stainless steel; elastomeric-switch membrane keyboard

Product: Modem and chronograph
Design Firm: Lee Payne Associates, Inc.
Atlanta, Georgia
Client: Hayes Microcomputer Products, Inc.
Norcross, Georgia
Materials: Extruded aluminum housing with snap-on injection-molded BAS end caps; translucent polycarbonate front plates permit LEDs to show through, whereas back plates are aluminum for grounding purposes

Product: Micropad
Design Firm: BIB Design Consultants
London, England
Client: Quest Automation
Dorset, England

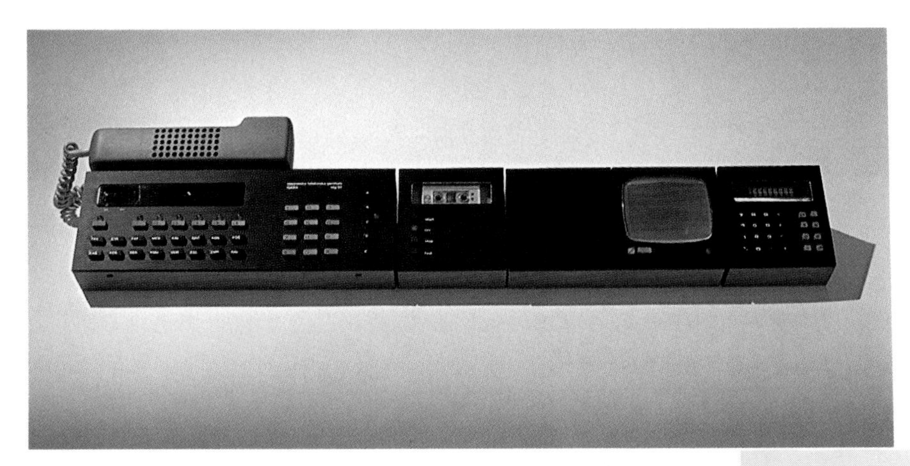

Product: Isicom Super
Designer: Davorin Savnik
Ljubljana, Yugoslavia
Client: Iskra Commerce
Ljubljana, Yugoslavia
Materials: Handset: aluminum; receiver: ABS plastic

Product: Nokia personal computer and workstation
Design Firm: Ergonomia Design
Turku, Finland
Client: Nokia Corporation
Helsinki, Finland
Awards: Seal of "Good Industrial Design," 1983 Hanover Exhibition

Product: Protocol convertor
Design Firm: Lee Payne Associates, Inc.
Atlanta, Georgia
Client: Technical Analysis Corporation
Atlanta, Georgia
Materials: Noryl structural foam

CHAPTER 6

NCESHOUSEWARESTOOLSAPPLIANCESHOUSEWARESTOOLSAPPLIANCESHOUSEWARESTOOLSAPPLIANCESHOUSEWARES
NCESHOUSEWARESTOOLSAPPLIANCESHOUSEWARESTOOLSAPPLIANCESHOUSEWAREST

Medical

ECTRONICSENTERTAINMENTHOMEELECTRONICSENTERTAINMENTHOMEELECTRONICS
AINMENTHOMEELECTRONICSENTERTAINMENTHOMEELECTRONICSENTERTAINMENTHOM

NGLIGHTINGLIGHTINGLIGHTINGLIGHTINGLIGHTINGLIGHTINGLIGHTINGLIGHTINGLIGHTINGLIGHTINGLIGHTING
NGLIGHTINGLIGHTINGLIGHTINGLIGHTINGLIGHTINGLIGHTINGLIGHTINGL

Equipment

CTRESIDENTIALFURNISHINGSCONTRACTRESIDENTIALFURNISHINGSCONTR
NTIALFURNISHINGSCONTRACTRESIDENTIALFURNISHINGSCONTRACTRESID

SSEQUIPMENTBUSINESSEQUIPMENTBUSINESSEQUIPMENTBUSINESSEQUIPMENTBUSINESSEQUIPMENTBUSINESSEQU
SSEQUIPMENTBUSINESSEQUIPMENTBUSINESSEQUIPMENTBUSINESSEQUIPMENTBUSINESSEQUIPMENTBUSINESSEQU

LEQUIPMENTMEDICALEQUIPMENTMEDICALEQUIPMENTMEDICALEQUIPMENTMEDICALEQUIPMENTMEDICALEQUIPMENT
ENTMEDICALEQUIPMENTMEDICALEQUIPMENTMEDICALEQUIPMENTMEDICALEQUIPMENTMEDICALEQUIPMENTMEDICAL

RIALEQUIPMENTTRANSPORTATIONINDUSTRIALEQUIPMENTTRANSPORTATIONINDUSTRIALEQUIPMENTTRANSPORTAT
RIALEQUIPMENTTRANSPORTATIONINDUSTRIALEQUIPMENTTRANSPORTATIONINDUSTRIALEQUIPMENTTRANSPORTAT

TIONALSPORTSEQUIPMENTRECREATIONALSPORTSEQUIPMENTRECREATIONALSPORTSEQUIPMENTRECREATIONALSPO
ENTRECREATIONALSPORTSEQUIPMENTRECREATIONALSPORTSEQUIPMENTRECREATIONALSPORTSEQUIPMENTRECREA

ESTEXTILESTEXTILESTEXTILESTEXTILESTEXTILESTEXTILESTEXTILESTEXTILESTEXTILESTEXTILESTEXTILES
ESTEXTILESTEXTILESTEXTILESTEXTILESTEXTILESTEXTILESTEXTILESTEXTILESTEXTILESTEXTILESTEXTILES

SFORTHEHANDICAPPEDDESIGNSFORTHEHANDICAPPEDDESIGNSFORTHEHANDICAPPEDDESIGNSFORTHEHANDICAPPED
HANDICAPPEDDESIGNSFORTHEHANDICAPPEDDESIGNSFORTHEHANDICAPPEDDESIGNSFORTHEHANDICAPPEDDESIGNS

Until recently, aesthetic design development of medical equipment has lagged far behind the sophisticated "state-of-the-art" technology which the equipment houses. Although this is no longer always the case, the delayed aesthetic development suggests that the design of medical equipment is an area in which the state of change continues to invite innovation.

What most signals the raised design consciousness of contemporary medical equipment is the attention paid to human factors engineering. While the role of human factors in design has certainly grown over the past ten years, nowhere is it so vital, or in such demand, as in the medical environment. It is, of course, crucial that an office chair comfortably support the human frame. Yet the comfort provided by a CT scanner detecting the brain tissues of an accident victim, and the convenience with which it can be operated by the technician and monitored by the radiologist in an emergency situation are surely on a different level. Ergonomics has a different meaning altogether in the life-and-death situations of a hospital.

What distinguishes medical equipment even more is the range of users it must address—the doctors and technicians who operate the equipment; the patients; and often, the service personnel. That medical products from dental equipment to nuclear magnetic resonance scanning systems are being designed to consider the entire range of users indeed sets them apart from their predecessors.

One product that first addressed these multivaried needs was the CT (computerized tomography) scanner, which captures a cross-sectional image of the human anatomy with X-rays, and then reconstructs the image electronically. With the development of its technology in the mid-1970s came the recognition that human factors would be integral to its application. The GE 9800 unit responds in a number of ways: The aperture of the gantry—the housing unit for the tube and detector—is larger than that of early models, and is thus less intimidating to the patient; softened edges of the gantry and the table both lessen the perceived volume and make for a more "human" appearance. The equipment rests on a small dark base. This gives it the illusion of floating, making it appear lighter and further diminishing its potentially threatening appearance. Finally, the scanner computers are located in a separate room—in part because their humming noise can disturb patients. As the designer for the scanner stated, it was the industrial design analysis "of user needs and CAD patient positioning study that established the scan geometry and patient positioning parameters permitting maximum tunnel opening, gantry tilt, and patient travel for improved patient access and scannability."

The participation of industrial designers in the development of NMR (nuclear magnetic resonance) scanners is even greater, as the unit makes heavier human factors demands. NMR collects chemical information of atoms by using the magnetic qualities of their nuclei. The NMR units, however, must generate a magnetic field thousands of times greater than the earth's, and as a result, some weigh over five tons. The problems confronting designers are: maintaining a uniform magnetic field by keeping the housing of the equipment, surrounding equipment, patients, and personnel free of ferrous materials; providing doctors and technicians with easy access to patients; and diminishing patient apprehension at being confined in a tube not much larger than his or her own body. Most manufacturers of these systems will agree that the human factors input of industrial designers and their collaboration with engineers has been crucial in resolving patient orientation, the placement of display equipment, means of patient/operator contact, and metal detection for the surrounding area and personnel.

While CT and NMR systems must consider human factors on a rather expansive scale, the manner in which smaller pieces of equipment meets these demands is no less significant. The Corning 102 Printer is a case in point. Designed to be used with existing blood analysis instrumentation, the printer meets the lab's needs for a permanent, sequential record of daily test and calibration results. It was the industrial designer's input which suggested that controls traditionally located internally and factory preset be made "user accessible" on the rear of the unit. The result was also a "user-installable" printer, which eliminated the need for a field service engineer. Likewise, the Corning designers approached the design of the Digital pH meter "from the inside out, dealing with the operational and perceptual issues before the structural ones." The comprehensive approach resulted in a design that, through a series of questions and statements, guides the user—who in all likelihood has had little or no previous exposure to the instrument—though the measurement sequence. The designers also specified that because a large LCD would "perform poorly in the extreme temperatures, vibrations, and shocks of industrial environments," the display medium be divided into two components—a standard seven-segment LED for numeric information, and a sixteen-segment alphanumeric LED for the parameters and instructions. A sealed touch panel, also geared to withstand the rigors of the industrial environment, reduced the number of key positions from 27 to 19, "simplifying operation and improving flexibility." A five-language capability for all display information was incorporated into the system to accommodate a world market. What all this points out most clearly is that the inside-out approach to design—by involving designers with systems engineers—results in a product whose fine-tuned aesthetics, coherent product image, ease of operation, and human factors recognition all correspond to the producer's high level of technology.

Finally, in discussing the user convenience of medical equipment, an irony becomes apparent: While the equipment must appear accessible and friendly, it should not necessarily appear *too* friendly or *too* accessible. The medical procedures carried out by this equipment are presumably of a serious nature, and to undermine this would also be to undermine the confidence of the patient. The equipment must be absolutely unintimidating, but it must also simultaneously convey a certain seriousness that will emphasize its capability and competence, thus reassuring the patient. While paradox may seem to prevail here, designers whose products can assimilate and balance contradictions are perhaps best recognizing the complex frame of mind of their users, and the complex situations of their use.

That the technology supporting most of these products is so recent, and in such a state of evolution, works both for and against the design of the equipment. Evolving products run the risk of being outdated even by the time they are introduced on the market; on the other hand, the opportunity exists to establish standards in an area of design that remains relatively uncharted. Designers who acknowledge the dual aspects of this situation will recognize the precarious position held by the design of their products. By acting on it, they will be contributing to the area of design that perhaps most pleads for fine tuning.

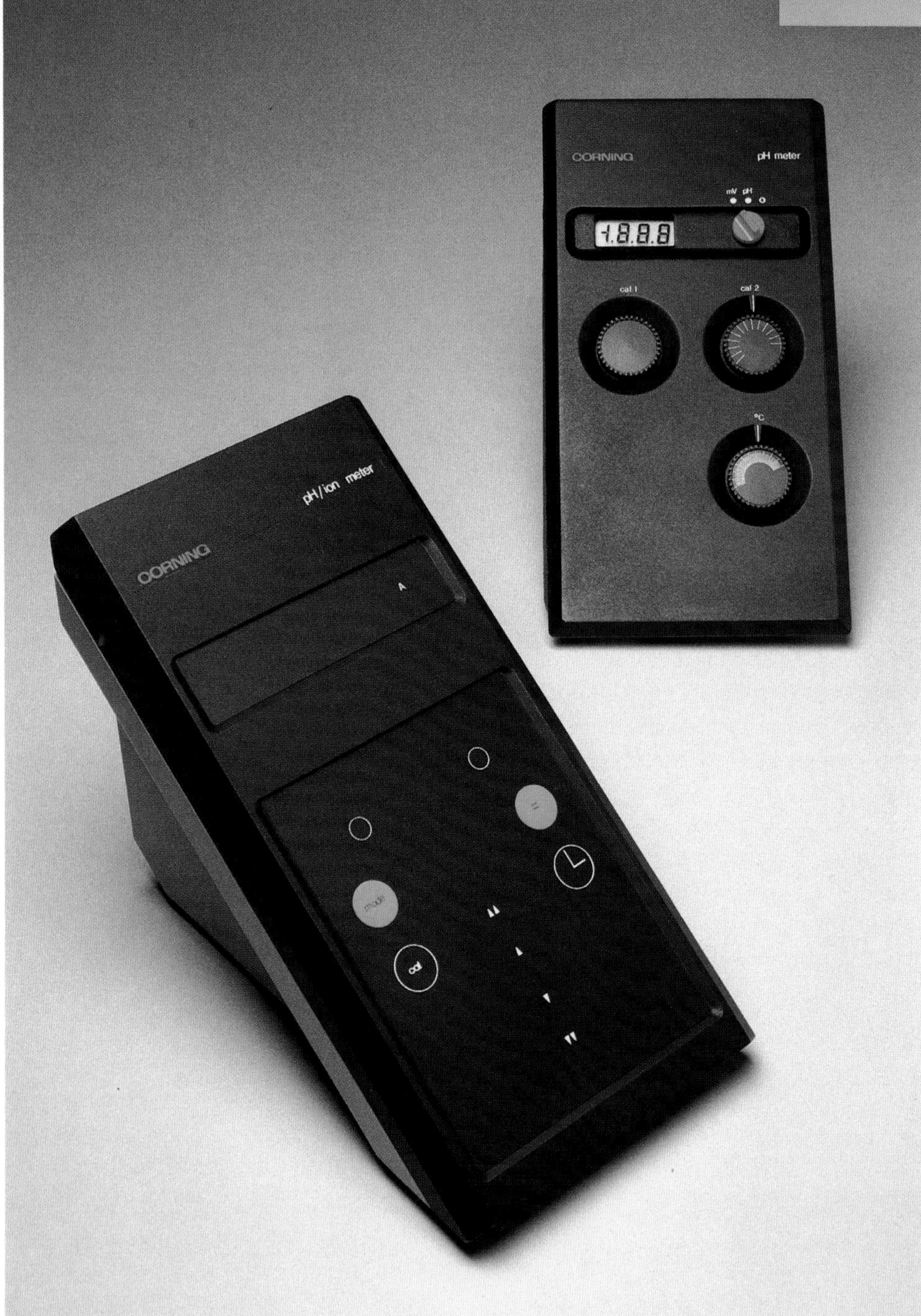

Product:	Digital pH meter
Designers:	Scott Mathis and Robert Potts Industrial Design Dept., Corning Medical Medfield, Massachusetts
Client:	Corning Science Products Medfield, Massachusetts
Awards:	1983 IDSA Industrial Design Excellence Award
Materials:	Molded Noryl foam; textured polyurethane finish; power supply chassis: pressure die-cast in aluminum; embossed touchpanel and display: fabricated of PVC; electrode arm and base: injection-molded modified polypropylene

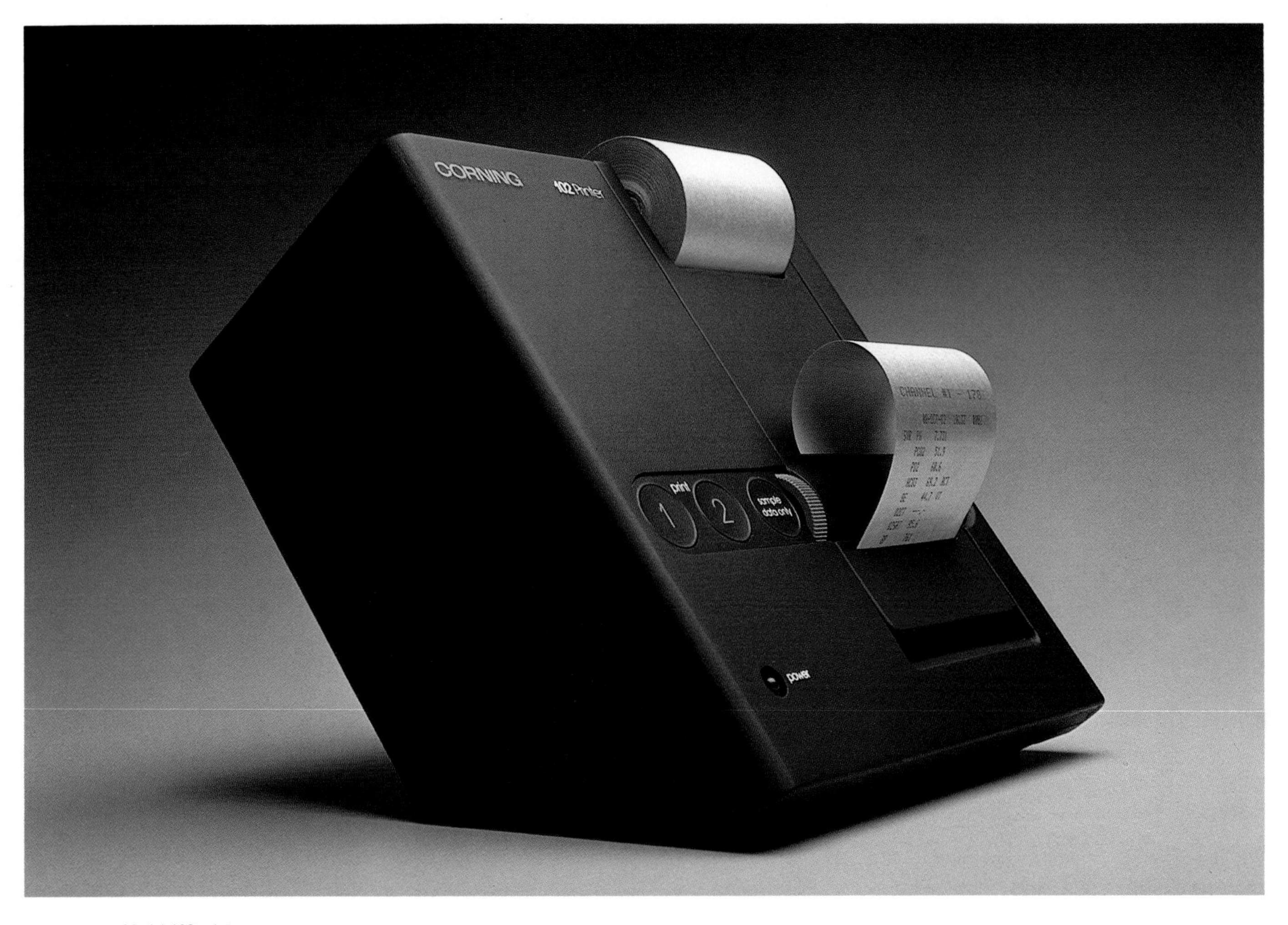

Product: Model 102 printer
Designer: John F. Buchholz, Corning Medical
Medfield, Massachusetts
Client: Corning Medical
Medfield, Massachusetts
Awards: 1983 IDSA Industrial Design Excellence Award
Materials: Noryl structural foam front housing with sonically welded boss inserts; back panel: thermoformed flame retardent ABS; silkscreened polycarbonate label provides interchangeable graphics on rear panel

Product: Driv-Lok Grooved Pin in aortic punch
Designer: Staff design, Driv-Lok, Inc.
Sycamore, Illinois
Client: Driv-Lok, Inc.
Sycamore, Illinois
Materials: Stainless steel

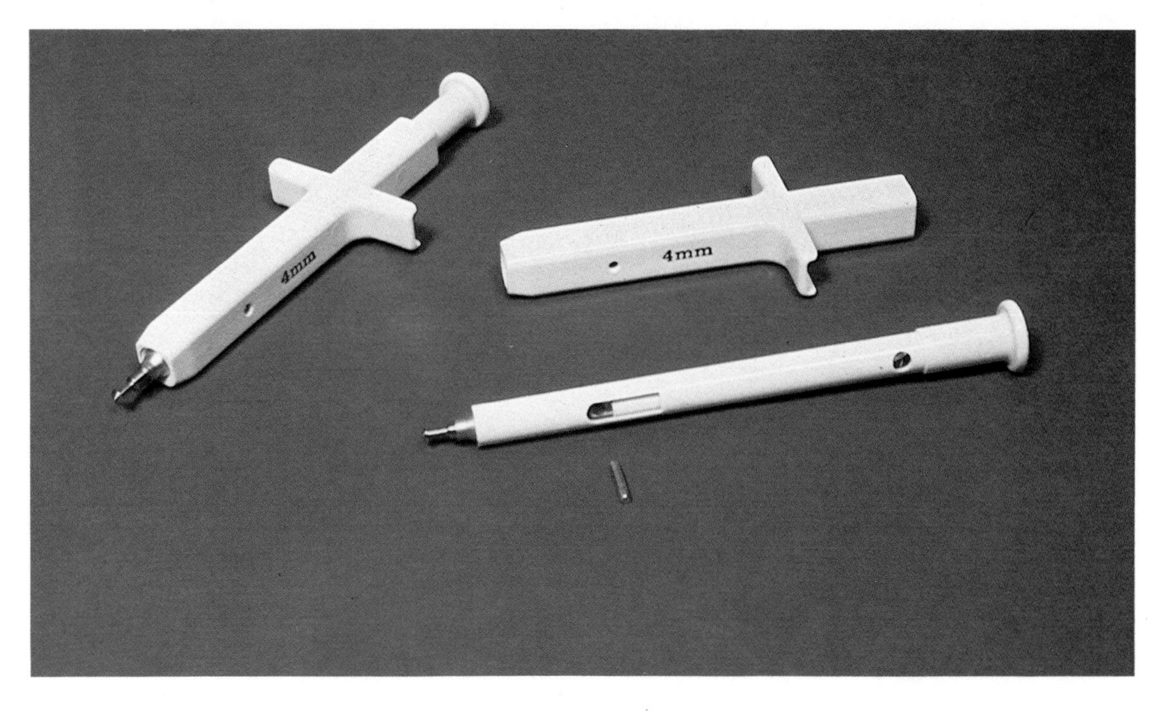

Product: Hypo-Count II blood glucose monitor
Designers: Staff design, Hypoguard Ltd. and Bernard Sams
Suffolk, England
Client: Hypoguard Ltd.
Suffolk, England
Awards: 1983 Design Council Award

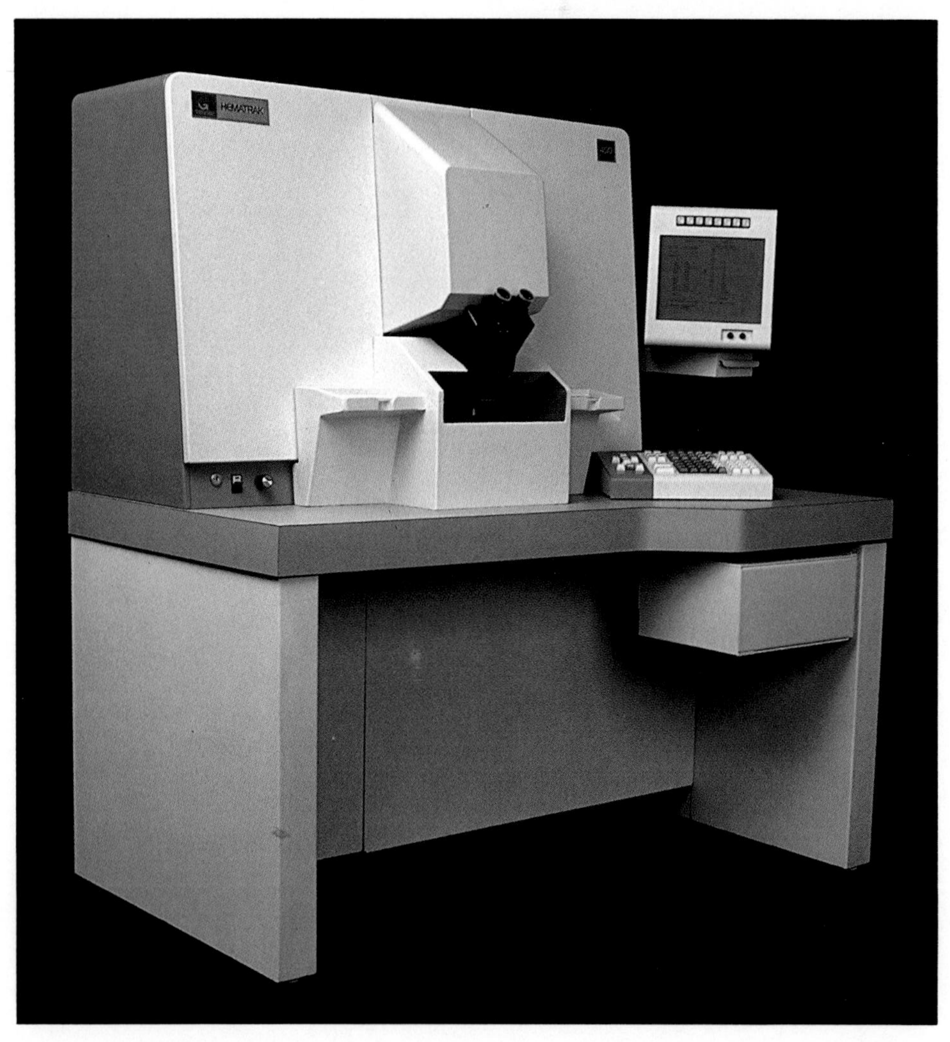

Product: Hematrak™, Geometric Data Corporation
Designer: William Sklaroff
Design Firm: William Sklaroff Design Associates
Philadelphia, Pennsylvania
Client: Geometric Data Corporation
division of SmithKlein Beckman
Wayne, Pennsylvania
Materials: Extruded aluminum; ABS plastic housing

Product: Electrophoresis densitometer
Designers: Gianfranco Zaccai, Instrumentation Laboratory Inc.
Lexington, Massachusetts
Client: Instrumentation Laboratory Inc.
Lexington, Massachusetts
Materials: Housing: Noryl structural foam; polyurethane painted finish

Product: Electronic Thermometer with LCD Readout
Designers: Engineering staff, Matsushita Electrical and Industrial Company of Japan
Osaka, Japan
Client: Panasonic Company
Division of Matsushita Electric Corporation of America
Secaucus, New Jersey
Materials: ABS plastic

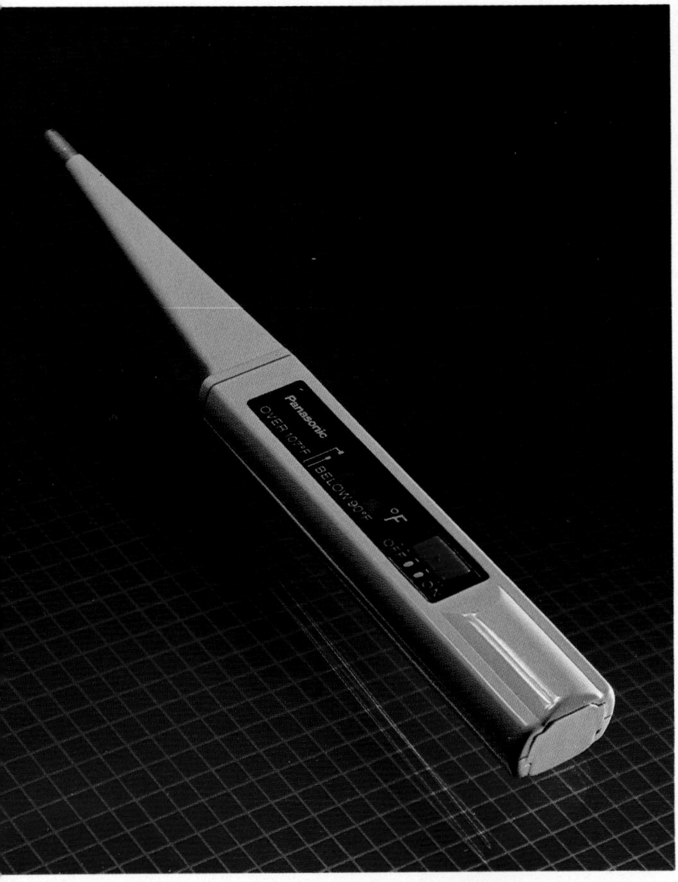

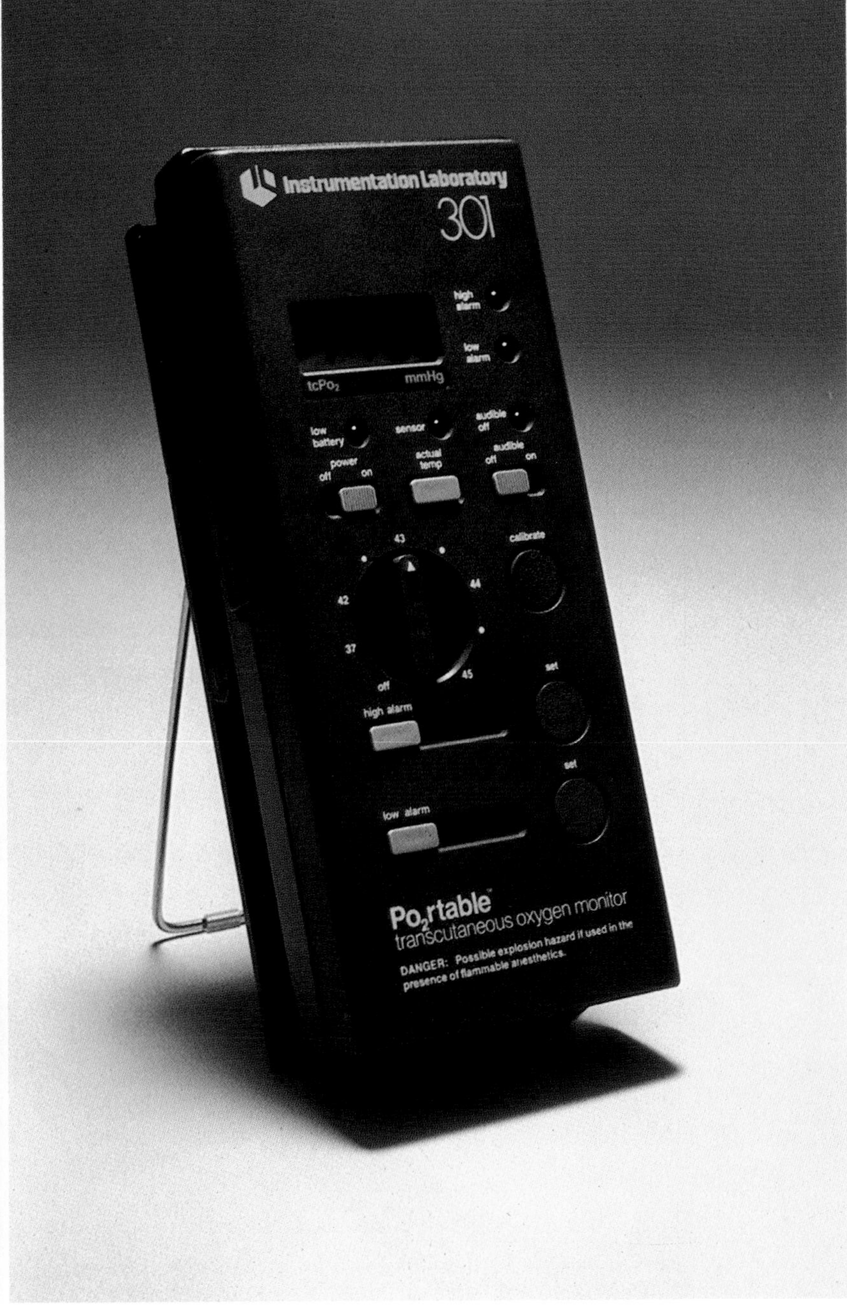

Product: PO_2rtable Oxygen Monitor
Designers: Staff design, Instrumentation Laboratory
Lexington, Massachusetts
Client: Instrumentation Laboratory, Inc.
Lexington, Massachusetts
Awards: *Industrial Design* magazine
1983 Design Review selection
Materials: Exterior: injection-molded in flame retardant cycolac ABS; kickstand: stainless steel

Product:	Fonar Beta 3000 Permanent Magnet Scanning System
Client:	Fonar Corporation Melville, New York
Materials:	Fiberglass exterior

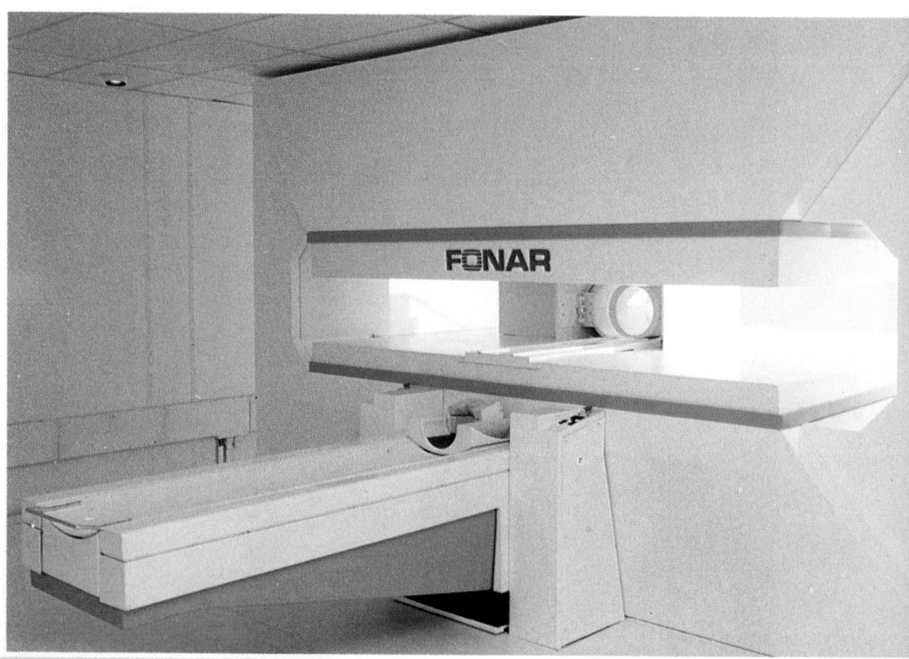

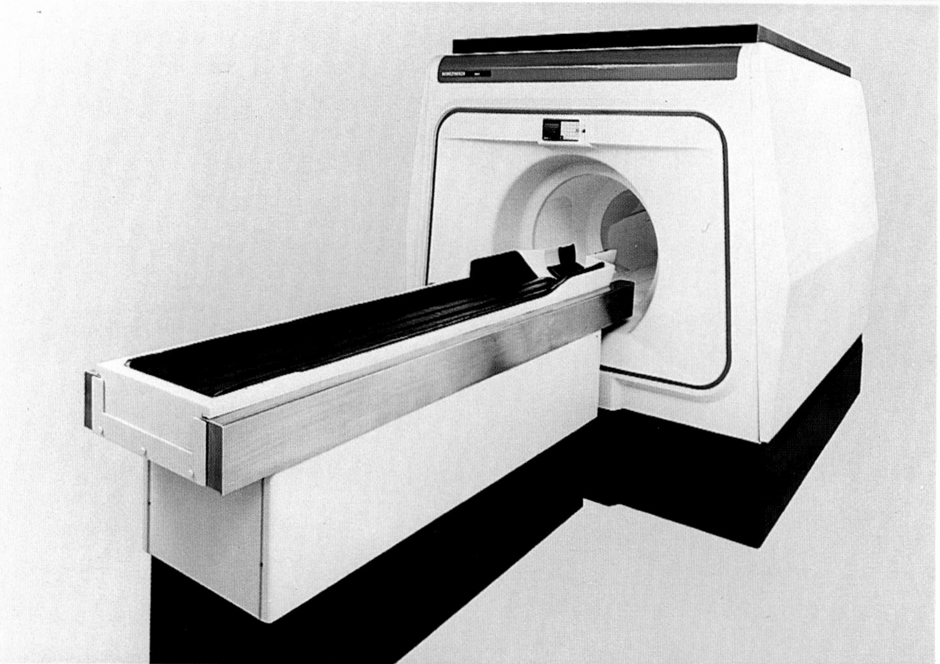

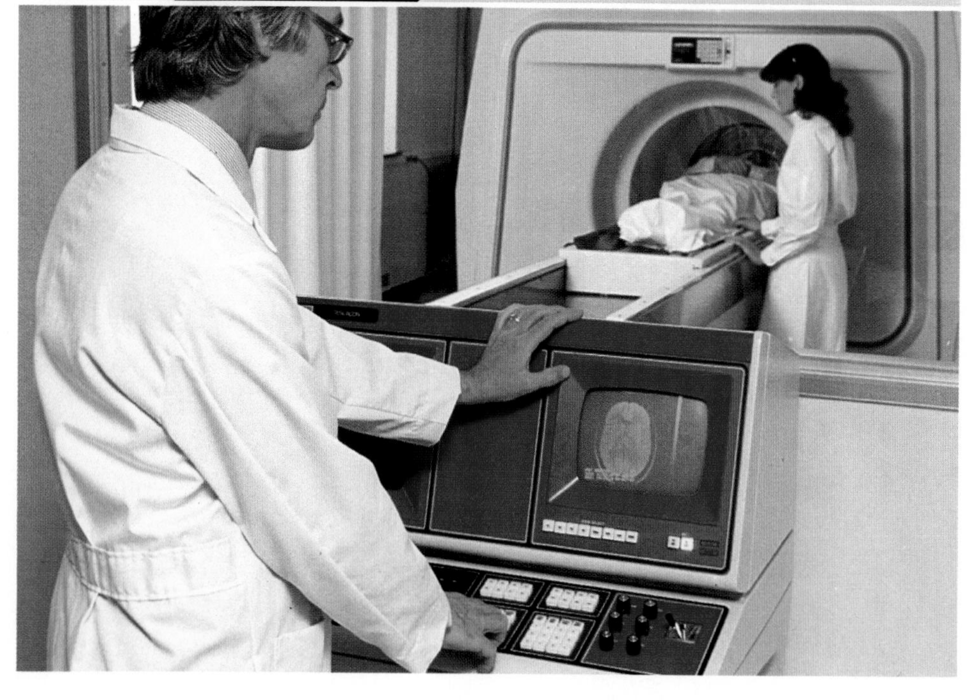

Product:	NMR Teslacon™ System
Designer:	Technicare Corporation Cleveland, Ohio
Client:	Technicare Corporation Cleveland, Ohio
Materials:	Fiberglass housing

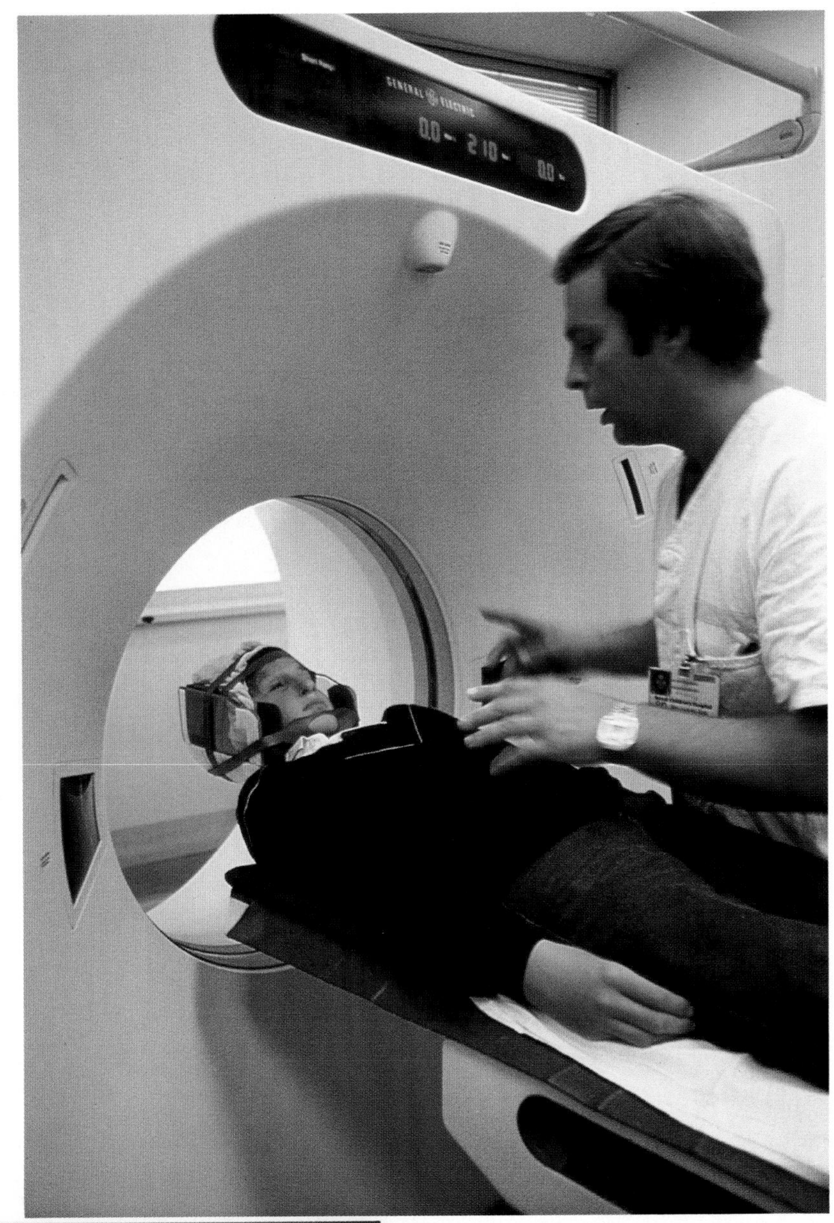

Product: CT 9800 computed tomography system
Designers: Staff Design
General Electric Medical Systems Operations
Milwaukee, Wisconsin
Client: General Electric Company, Medical Systems Operations
Milwaukee, Wisconsin
Awards: 1982 IDSA Design Excellence Award
Materials: Fiberglass for form delineation of gantry and table pedestal covers; structural foam with sprayed metal epoxy tooling on moveable cradle carriage and console keyboard; vacuum-formed monitor display materials; self-skinning foam surface with extended edges

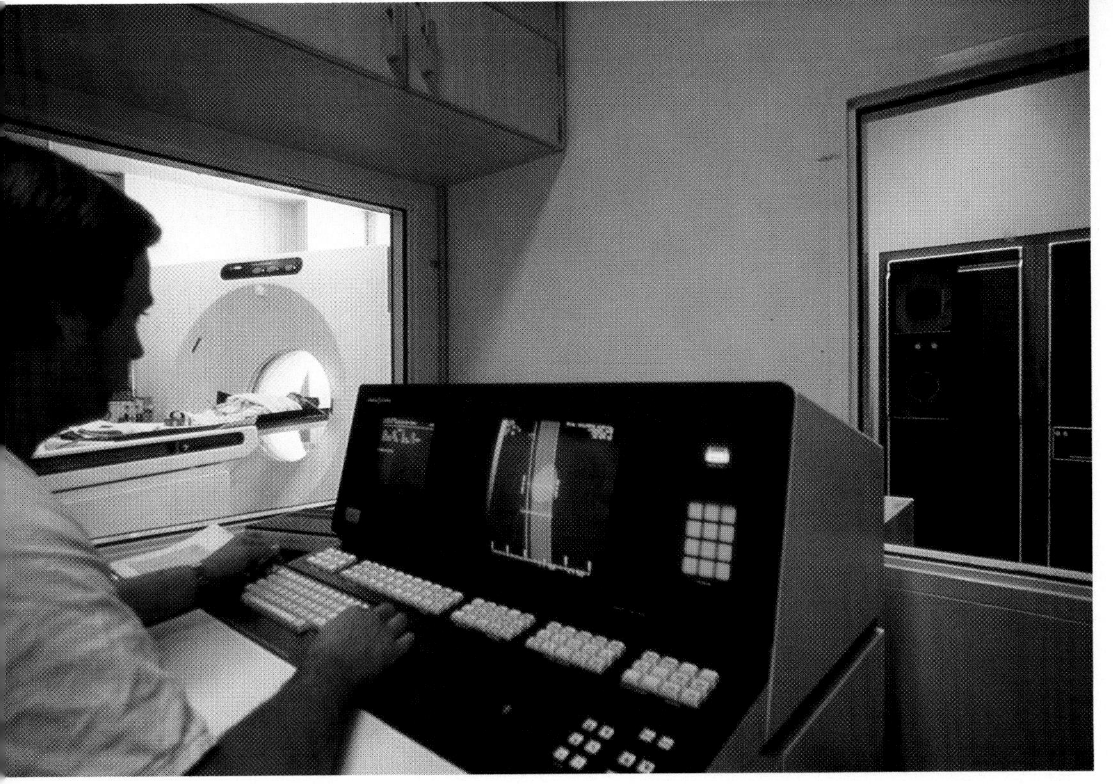

Product: Telemate
Design Firm: Coons and Beirise Design Associates
Cincinnati, Ohio
and staff design, Hill-Rom Company
Batesville, Indiana
Client: Hill-Rom Company
Batesville, Indiana
Awards: *Industrial Design* magazine
1983 Design Review selection
Materials: Handset and holster: injection-molded, high-impact ABS; holster-release latches: injection-molded polycarbonate; handset keyboard: molded silicon rubber; graphic backplate: silkscreened adhesive-back aluminum insert

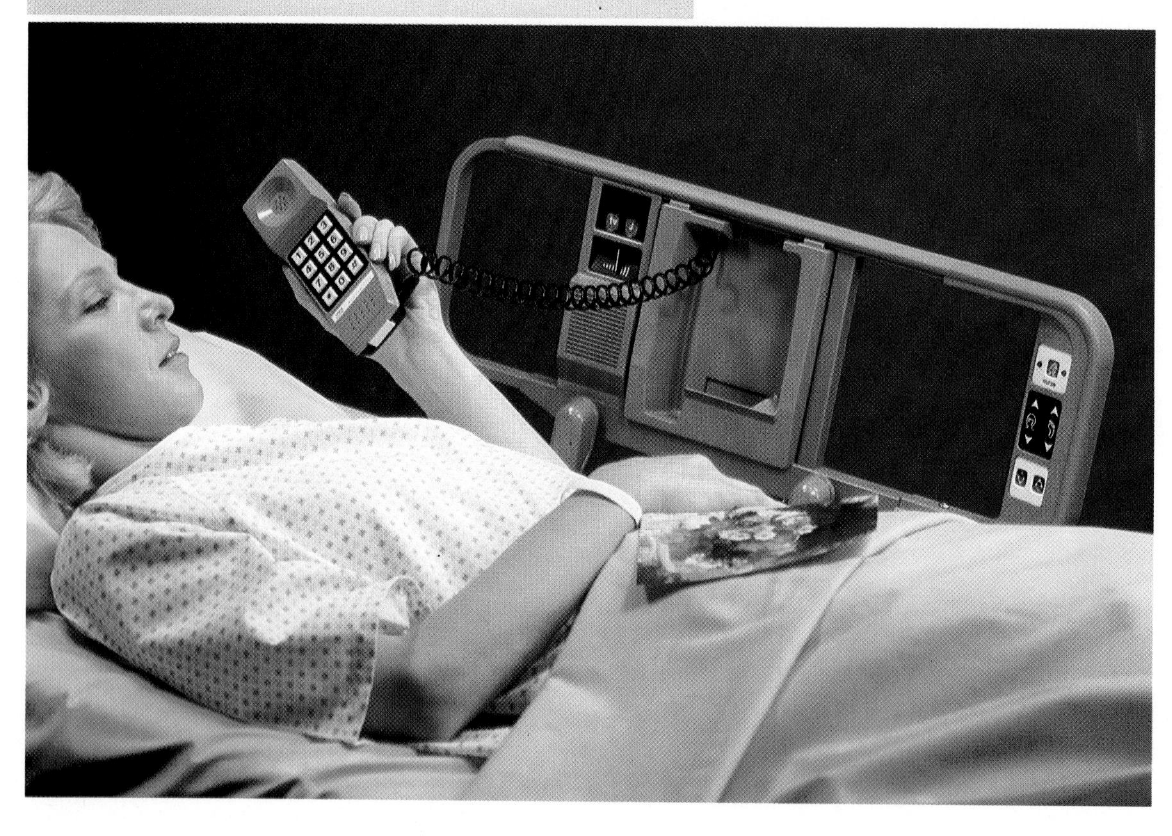

Product: 943 Flame Photometer
Designer: Gianfranco Zaccai
Boston, Massachusetts
Client: Instrumentation Laboratory SpA
Milan, Italy
Awards: 1982 *Industrial Design* magazine
Design Review selection
1983 IDSA Industrial Design Excellence
Award
Materials: Housing: sheet aluminum and structural foam with polyurethane finish; membrane keyboard/display control panel: polycarbonate

Product: Pillow Speaker
(patient communications signalling device)
Designer: William Sklaroff
Design firm: William Sklaroff Design Associates
Philadelphia, Pennsylvania
Client: Executone Inc.
Jericho, New York
Materials: ABS "cycolac" housing

Product: Pherotron modular densitometer system
Design Firm: Neumeister Design
Munich, West Germany
Client: LRE Medizintechnik
Munich, West Germany

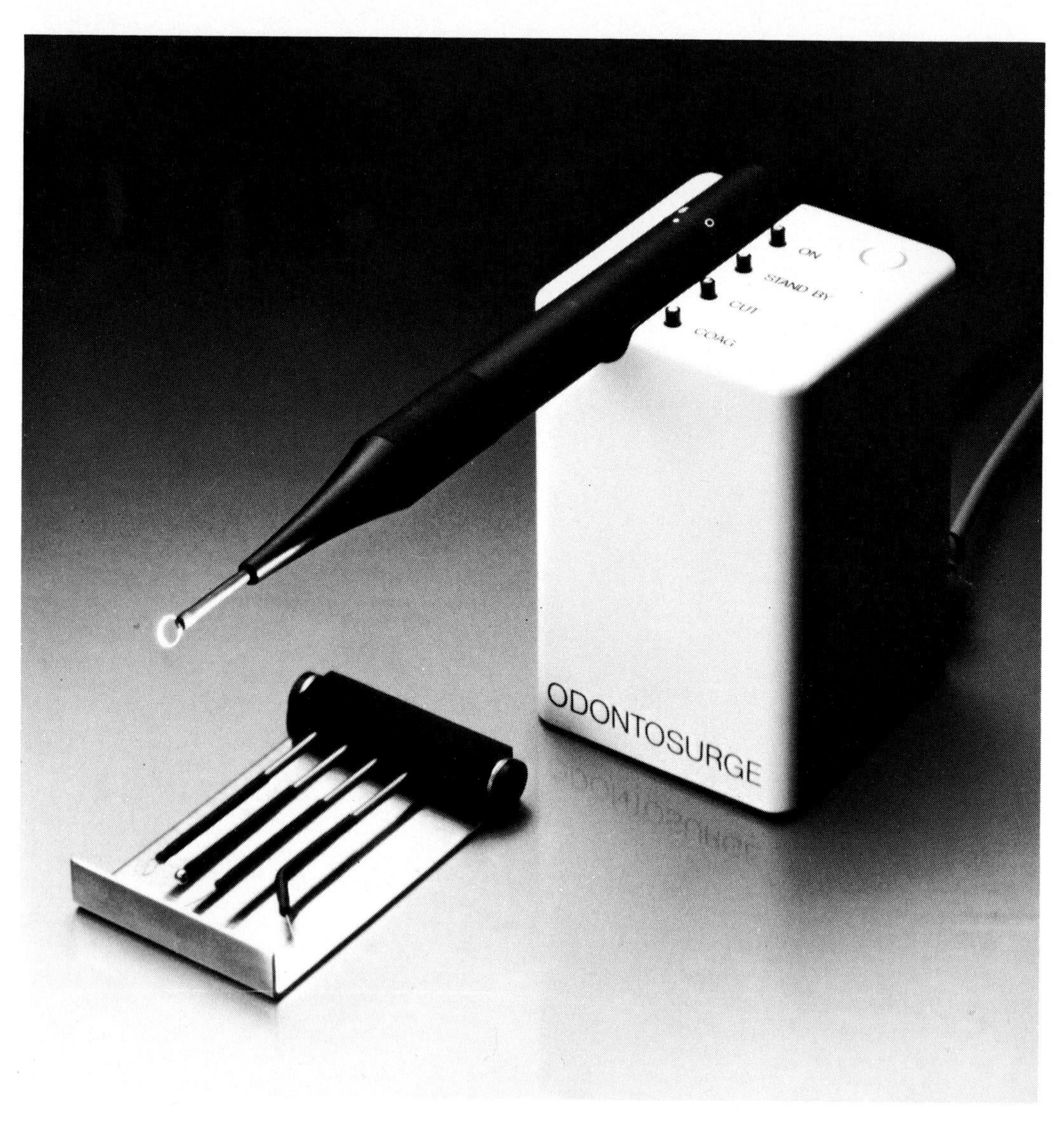

Product: Odontosurge
Designer: David Lewis and Lennart Goof
Hørsholm, Denmark
Client: A/S L. Goof
Hørsholm, Denmark
Awards: 1982 Danish Design Council Award

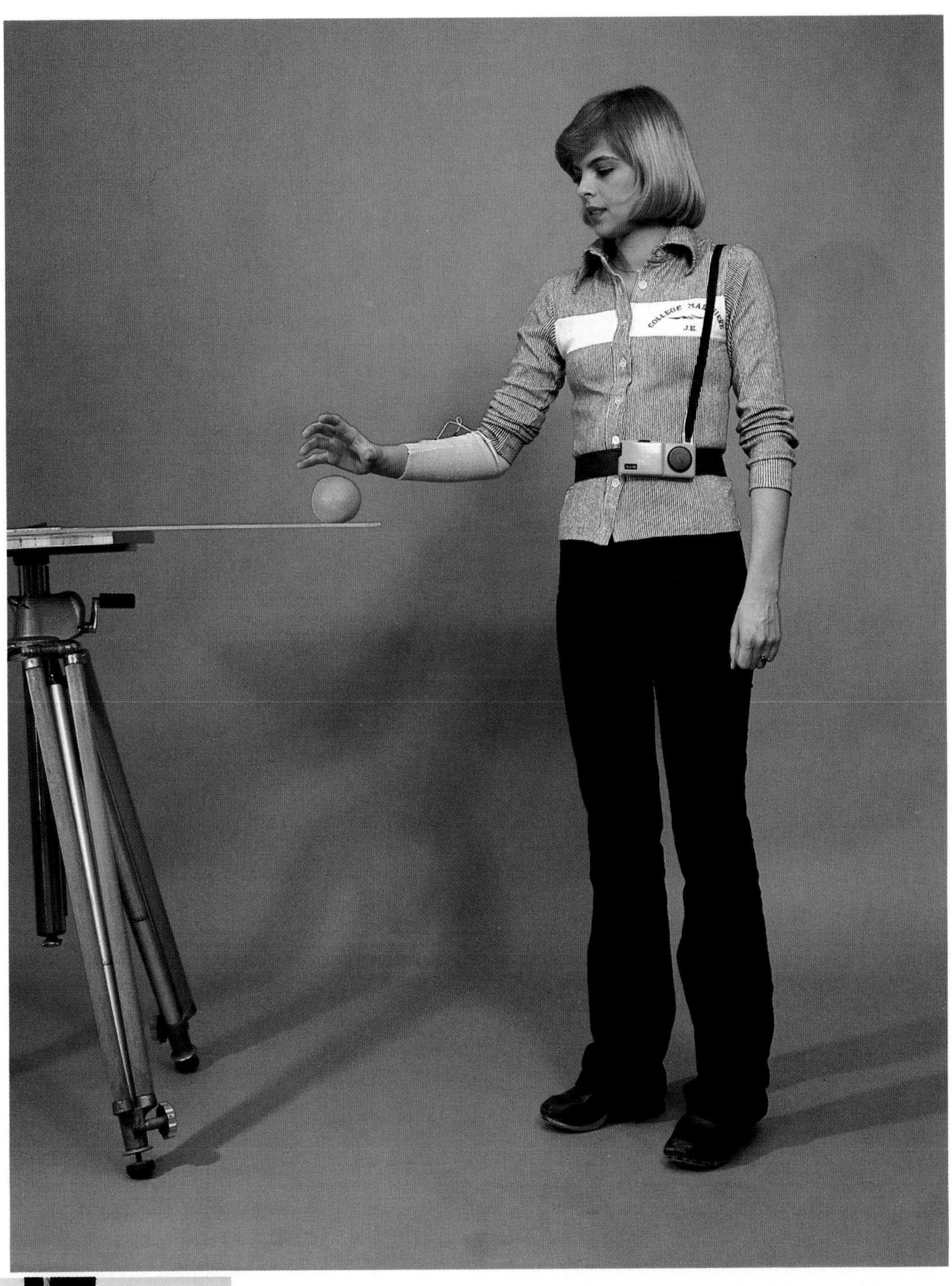

Product:	Hand Prehension Stimulator FESE H3
Designer:	Davorin Savnik Ljubljana, Yugoslavia
Client:	Institut J. Stefan Ljubljana, Yugoslavia
Awards:	Stuttgart Design Centre Award Biennial of Industrial Design BIO 7 Gold Medal, Ljubljana Design Centre Belgrade Award
Materials:	Molded ABS plastic

Product: AudioScope™
Designers: Paul Sweeney, Arthur Pulos, Eric Beyer, and Matthew Murray
Pulos Design Associates, Inc.
Syracuse, New York
Design Firm: Pulos Design Associates, Inc.
Syracuse, New York
Client: Welch Allyn, Incorporated
Skaneateles Falls, New York
Materials: Housing: ABS plastic; halogen lamp; rechargeable nickel cadmium battery

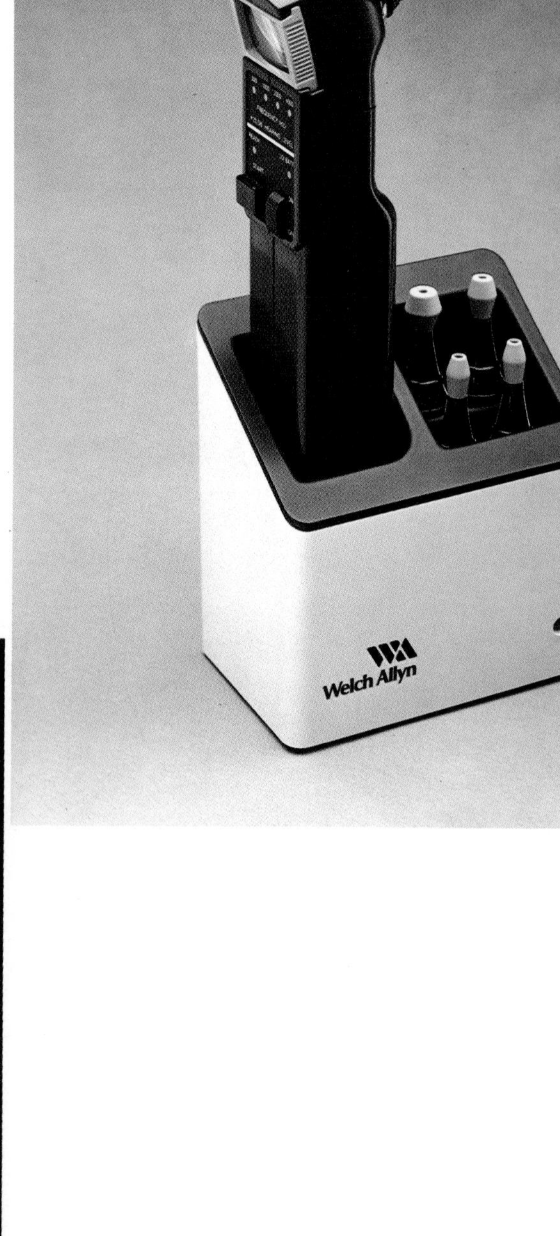

Product: Cardiac Strobe
Designer: Oskar Heininger
Framingham, Massachusetts
Client: Cardiac Imaging Inc.
Medfield, Massachusetts
Awards: 1983 IDSA Industrial Design Excellence Award
Materials: Aluminum with epoxy finish

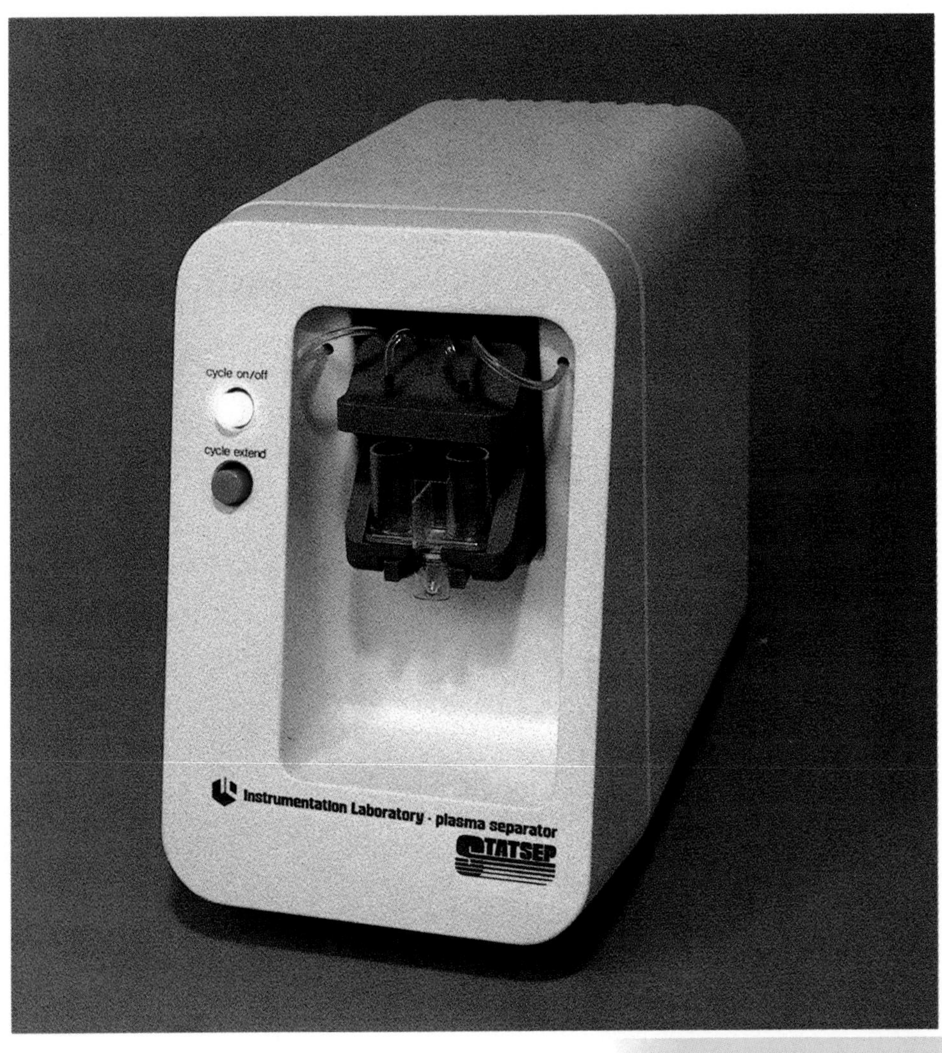

Product:	Statsep plasma separator
Designer:	Gianfranco Zaccai Boston, Massachusetts
Client:	Instrumentation Laboratory Inc. Lexington, Massachusetts
Awards:	1982 *Industrial Design* magazine Design Review selection
Materials:	RIM Mobay Baydur 724 for front and rear housing sections; sheet metal for subassembly bracket inside front panel; cast urethane for clamping pressurizing jaws

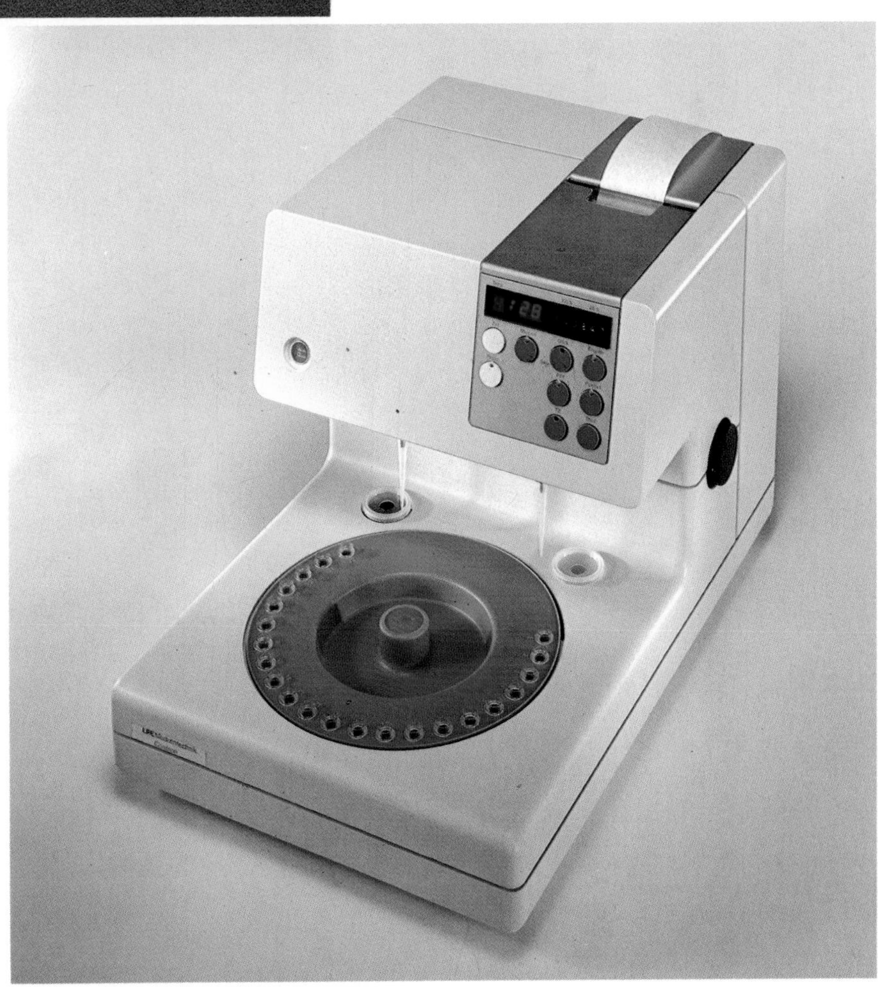

Product:	Coatron coagulometer
Design Firm:	Neumeister Design Munich, West Germany
Client:	LRE Medizintechnik Munich, West Germany

CHAPTER 7

CESHOUSEWARESTOOLSAPPLIANCESHOUSEWARESTOOLSAPPLIANCESHOUSE
CESHOUSEWARESTOOLSAPPLIANCESHOUSEWARESTOOLSAPPLIANCESHOUSE

CTRONICSENTERTAINMENTHOMEELECTRONICSENTERTAINMENTHOMEELECT
INMENTHOMEELECTRONICSENTERTAINMENTHOMEELECTRONICSENTERTAIN

GLIGHTINGLIGHTINGLIGHTINGLIGHTINGLIGHTINGLIGHTINGLIGHTINGLIGHTINGLIGHTINGLIGHTINGLIGHTING
GLIGHTINGLIGHTINGLIGHTINGLIGHTINGLIGHTINGLI

TRESIDENTIALFURNISHINGSCONTRACTRESIDENTIALF
TIALFURNISHINGSCONTRACTRESIDENTIALFURNISHING

SEQUIPMENTBUSINESSEQUIPMENTBUSINESSEQUIPMENTBUSINESSEQUIPMENTBUSINESSEQUIPMENTBUSINESSEQU
SEQUIPMENTBUSINESSEQUIPMENTBUSINESSEQUIPME

EQUIPMENTMEDICALEQUIPMENTMEDICALEQUIPMENTME
NTMEDICALEQUIPMENTMEDICALEQUIPMENTMEDICALE

Industrial Equipment and Transportation

RIALEQUIPMENTTRANSPORTATIONINDUSTRIALEQUIPMENTTRANSPORTATIONINDUSTRIALEQUIPMENTTRANSPORTAT
RIALEQUIPMENTTRANSPORTATIONINDUSTRIALEQUIPMENTTRANSPORTATIONINDUSTRIALEQUIPMENTTRANSPORTAT

IONALSPORTSEQUIPMENTRECREATIONALSPORTSEQUIPMENTRECREATIONALSPORTSEQUIPMENTRECREATIONALSPO
NTRECREATIONALSPORTSEQUIPMENTRECREATIONALSPORTSEQUIPMENTRECREATIONALSPORTSEQUIPMENTRECREA

STEXTILESTEXTILESTEXTILESTEXTILESTEXTILESTEXTILESTEXTILESTEXTILESTEXTILESTEXTILESTEXTILES
STEXTILESTEXTILESTEXTILESTEXTILESTEXTILESTEXTILESTEXTILESTEXTILESTEXTILESTEXTILESTEXTILES

FORTHEHANDICAPPEDDESIGNSFORTHEHANDICAPPEDDESIGNSFORTHEHANDICAPPEDDESIGNSFORTHEHANDICAPPED
ANDICAPPEDDESIGNSFORTHEHANDICAPPEDDESIGNSFORTHEHANDICAPPEDDESIGNSFORTHEHANDICAPPEDDESIGNS

What most distinguishes the design of transportation and heavy industry from other categories of design are the stringent demands made upon it by the environment. In most areas of design, the designer must contend with the product and the user alone. Here, however, he must also consider how the product addresses the environment. This is all the more relevant at a time in history when conservation has become a particularly sensitive, if not critical, issue.

Consider, for example, the Roulund floating oil boom, designed to stop the spread of oil slicks caused by leaks in drilling rigs and ships. The product is simply conceived and simply executed; modular rubber floats are large enough to prevent spillage, and have been designed to lie on the water's surface at an angle which will stabilize them when they are towed in a U-shape—enough to withstand collisions with ships. They are also easily cleaned and easily transported in containers, either by plane or truck to the accident site. Reflector strips identify inside and out. Both in concept and design, the product meets urgent contemporary demands of the environment in a straightforward and effective manner.

Magnetic levitation trains, such as the Transrapid 06 designed by Alexander Neumeister, show the same attention, though on a far grander scale. The train, designed along the principles of electromagnetic levitation, run along their tracks without any frictional contact. In essence, ferromagnetic armature rails of the track and electromagnets fitted to the train maneuver transit; levitation magnets pull the train up to the track while guidance magnets keep it centered. The elevated tracks demand minimal space and disruption of the countryside; the noise level and energy consumption are low and the electric drive system does not emit gases or fumes. Clearly, this is an example of a rapid transit system well-tuned to its environment. Whether it is a simply constructed oil boom or the new technology for rapid transit, there is a direct equation between the sophistication of the products in this category and their recognition of the affect they may have on an often fragile environment.

The aerodynamic styling of the Transrapid 06 is also evident elsewhere in transportation design. More emphatic streamlining for diminished air resistance and susceptibility to crosswinds are as apparent in automobile design. As Toyota states about its 1984 van, "The vehicle spent a lot of time in the wind tunnel during its early design stages." What resulted was an aerodynamically styled front end with a 40 degree rake to the windshield, and body lines that yield a low drag coefficient of .40 (which is lower than even some passenger cars). The sloping front and roofline also make for a downforce that increases stability at high speeds. Likewise, wind tunnel research contributed to the design of the 1984 Chevrolet Corvette. The airflow through its passenger compartment and engine was as thoughtfully considered as the airflow over the outside of the car. As the designer explains, "Air is admitted from under the front end rather than through the traditional grille; both drag and lift are minimized, as is wind, noise, and dirt contamination." The overall drag coefficient in this case is .341.

The Corvette's ergonomic interior is as unique as its aerodynamic exterior, another aspect appearing more frequently in contemporary automobile design. Ergonomically designed car seats consider two things—a form which gives the human frame maximum support, and a high degree of adjustability for individual contours. Although these features may not come as a surprise in the Corvette—which its designers refer to as "a driver's car rather than a passenger car . . . that must combine excellent roadholding with immediate response and a high degree of personal interaction for the driver . . ." —their appearance in more standard vehicles suggests that another mark of contemporary design is that ergonomics in the automobile is no longer limited to luxury cars.

Human factors are also making a strong showing in the control display designs of transportation vehicles. As the designers of the Corvette explain, "The controls are designed for easy reach, and liquid crystal digital displays deliver instant readings of engine condition and driving range." While automobiles are perhaps the most obvious arena for human factored controls, they are evident elsewhere as well. The Crown Series TS Turret Sideloader designed by Richardson/Smith is a case in point. The designers' research determined not only the placement of controls, but in some cases, generated the graphics as well. The panel for the sideloader includes controls for rotation, traverse, forward and reverse travel, a discharge indicator, emergency override switches, a horn button, and a lighted function display panel. The designers discovered through their research that a combination of established and new codes would most effectively indicate the full range of these functions. This, too, sets apart contemporary vehicles and transport equipment—the clear placement and graphics of controls which recognize human factors specifications. As products are designed more and more for international markets, this attention to graphics, and to universal signage in particular, will increase all the more.

Finally, in this category of design, the use of new materials cannot be overlooked. The Avtek 400, for example, a six- to nine-passenger twin turboprop plane, uses lightweight, high-strength composites of Kevlar aramid fiber and honeycomb of Nomex aramid fiber; the weight of the plane is 3,100 lbs. (1,406 kg.), roughly half that of similar aircraft. Kevlar—stronger than steel and aluminum but lighter than fiberglass—has rather obvious applications in aircraft construction. With a top speed of 425 miles (684 km.) per hour and the ability to reach an altitude of 10,000 feet (3048 m.) in one minute and fifty seconds, the plane's performance surpasses that of comparably sized aircraft. Again, it is the design process—here the application of new materials—which has revolutionized the product by so dramatically upgrading its performance.

The design of contemporary transportation and heavy industry then, is signalled by a number of features; most important, it tends to acknowledge the environment which is to accommodate it. Add to this more dramatic aerodynamic styling, ergonomic finetuning, and the application of new materials, and it becomes evident that the transitions occurring here render a new vitality for the design of this equipment.

Product:	1984 Chevrolet Corvette
Designers:	General Motors Design, Warren, Michigan
Client:	Chevrolet Motor Division General Motors Corporation Warren, Michigan
Awards:	1983 IDSA Industrial Design Excellence Award
Materials:	Body: fiberglass formed on matched dies with panels designed for easy repair; bumpers: elastomeric with protective molding along the body side; micro-electric computer system; four-wheel independent suspension with forged aluminum guide arms and glass fiber leaf springs tuned to specially designed tires

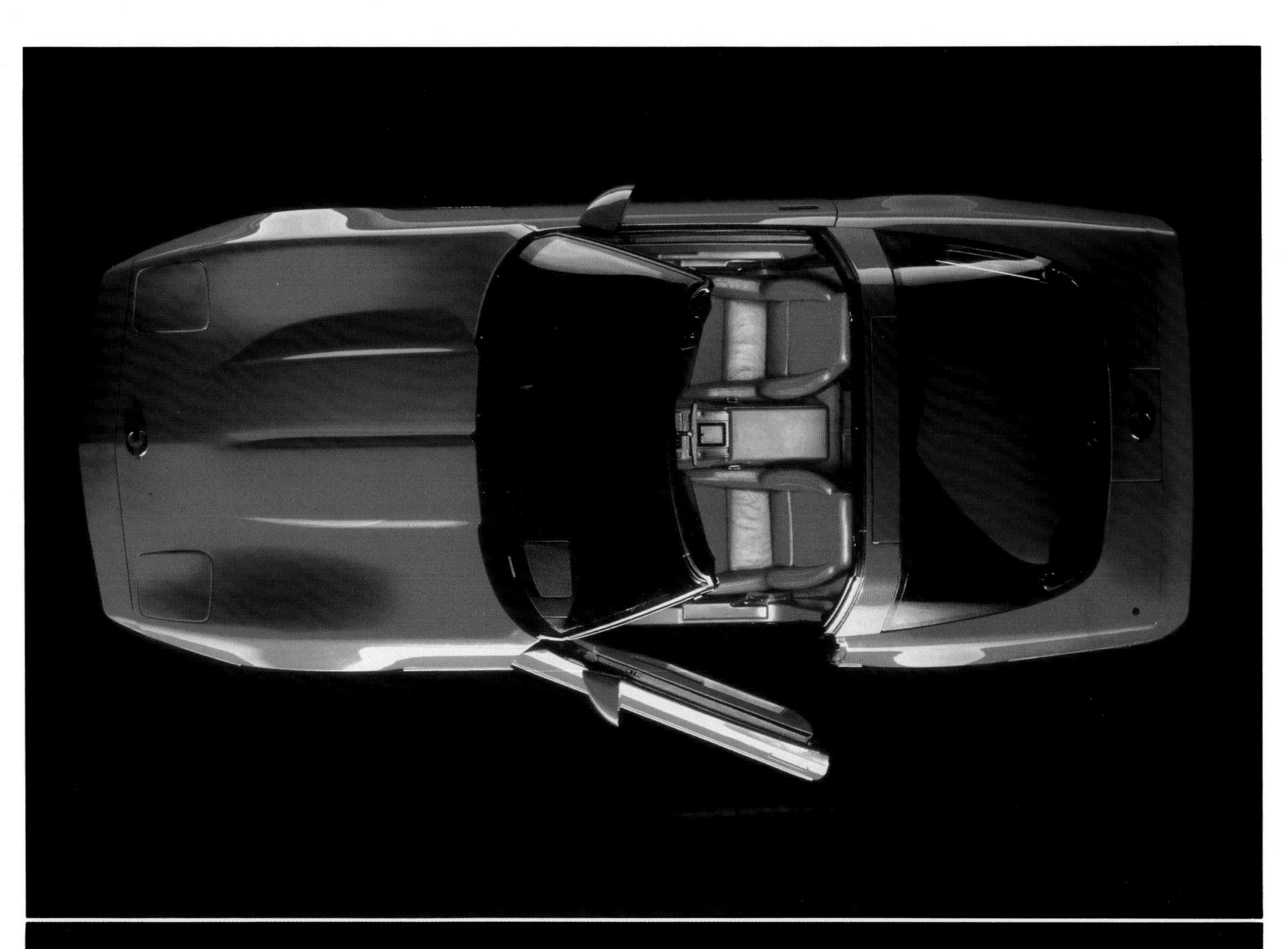

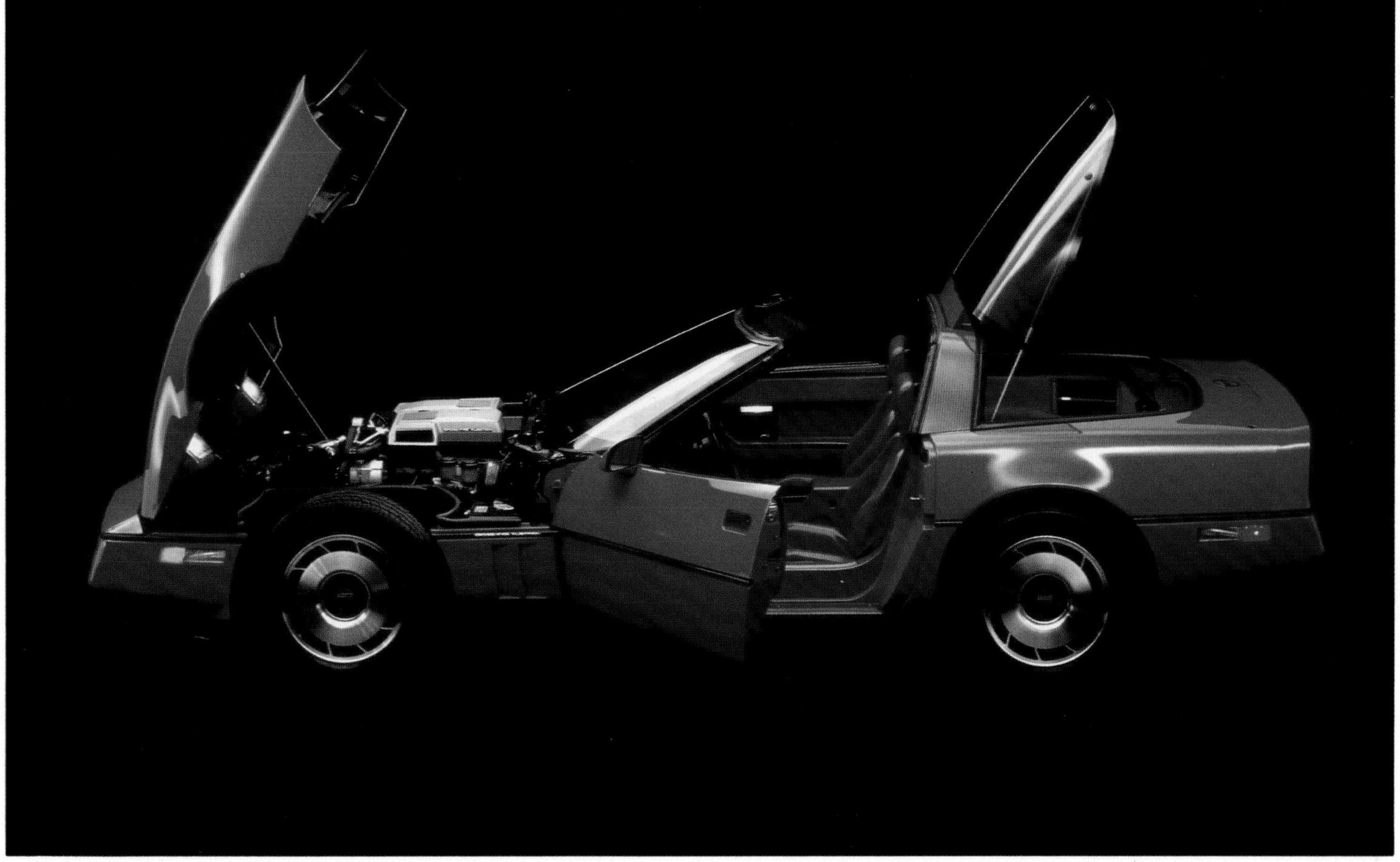

CORVETTE

Product:	1984 Nissan Stanza
Designers:	Design team, Nissan Motor Co. Ltd. Tokyo, Japan
Client:	Nissan Motor Co., Ltd. Tokyo, Japan
Materials:	Steel unibody

Product:	Ford Sierra
Designer:	Ford European product development team headed by Charles Knighton in conjunction with R. W. Mellor and Uwe Bahnsen
Client:	Ford of Europe Inc. Essex, England
Awards:	1983 Design Council Award

Product:	1984 Toyota Van
Designers:	Design staff, Toyota Motor Corp. Toyota City, Japan
Client:	Toyota Motor Corp. Toyota City, Japan
Materials:	Aerodynamically styled steel body and 40-degree rake to windshield yield drag coefficient of .40; electronic fuel injection; hydraulic valve lifters; maintenance-free battery; integrated ignition system; platinum-tipped spark plugs; 5-bearing crankshaft with eight balancers

Product:	LRC Train interior
Designer:	Morley Smith Montreal, Quebec, Canada
Design Firm:	GSM Design Inc. Montreal, Quebec, Canada
Client:	VIA Rail Canada Montreal, Quebec, Canada
Materials:	Seats: fiberglass-reinforced polyester; cushions: neoprene foam; upholstery: wool and nylon; floorcovering: nylon; baggage compartments: vacuum-formed ABS plastic; interior panels: melamine laminate

Product:	GSM Taxi
Designer:	Morley Smith Montreal, Quebec, Canada
Design Firm:	GSM Design Inc. Montreal, Quebec, Canada
Awards:	Design Canada Special Award 1982 for research and innovation
Materials:	Steel main frame; ιubular subframe; energy-absorbing perimetric bumper; lower body section: fiberglass moldings bolted to subframe; bumper and side panel system: urethane; interior lining: rigid plastic door and roof panels and window surrounds; taxi can accommodate wheelchair passenger

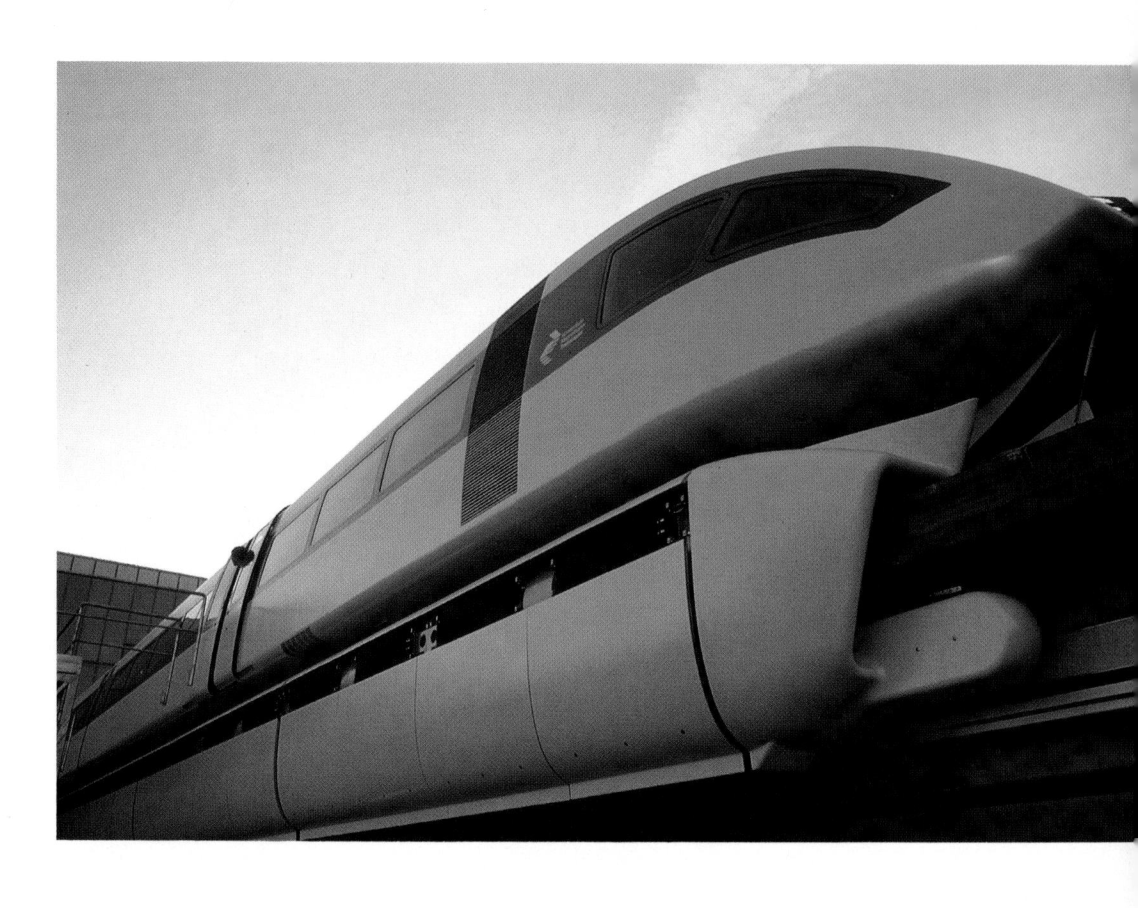

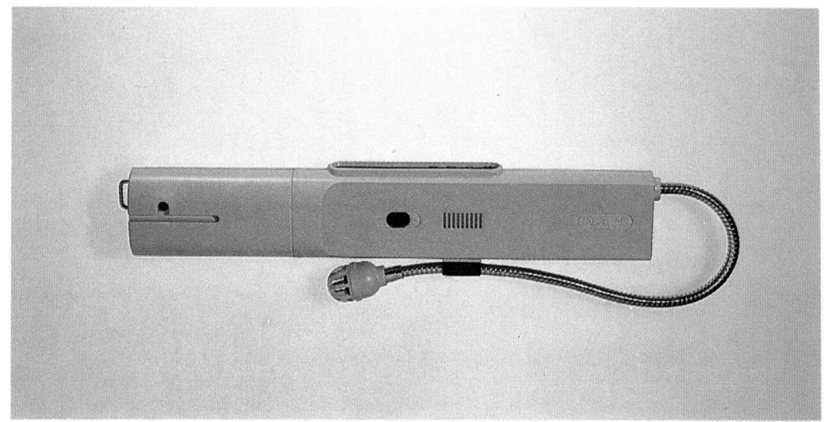

Product:	Gas-Trac combustible gas detector
Design Firm:	Dale E. Fahnstrom Design Riverside, Illinois
Client:	Epsilon Lambda Electronics Batavia, Illinois
Awards:	1981 *Industrial Design* magazine Design Review special mention
Materials:	Component cover, probe cover, and switch injection-molded ABS; probe clip and thumbwheel: black nylon; lanyard loop: stainless steel; battery chips: phosphor bronze; escutcheon plate: aluminum; gooseneck: flexible chrome-plated steel

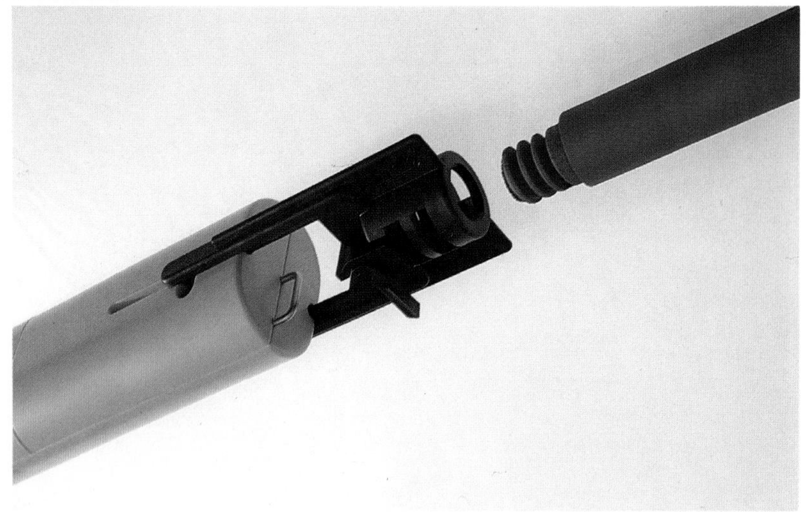

Product: Transrapid 06 magnetic levitation train
Designer: Alexander Neumeister
Munich, West Germany
Design Firm: Neumeister Design
Munich, West Germany

Product:	AVTEK 400
Designers:	William Taylor, Palm Desert, California Al Mooney, Ingram, Texas
Client:	Avtek Corporation Camarillo, California
Materials:	Airframe: honeycomb of "Nomex" aramid fiber sandwiched between skins of "Kevlar" aramid fiber

Product: Boeing 767 interior
Design Firm: Walter Dorwin Teague Associates
New York, New York
in collaboration with Boeing Engineers
Boeing Commercial Airplane Company
Seattle, Washington
Client: Boeing Commercial Airplane Company
Seattle, Washington
Materials: Compartments and interior hard surfaces: formed Nomex(TM) fire-retardant, crushed-core honeycomb materials; opaque Tedlar(TM) pre-printed or plain material silkscreened to specialized airline design; clear Tedlar protective film; fabrics: wool or wool blends manufactured to specifications of each airline

Product:	Westland 30 Series helicopter
Designers:	R. A. Doe, G. H. Humphrey, D.E.H. Balmford R. E. Swinfield, F. G. Rivers, P. Brammer, and B. Main
Client:	Westland Helicopters Ltd. Somerset, England
Awards:	The Duke of Edinburgh's Designer's Prize 1983 Design Council Award
Materials:	Twin Rolls-Royce Gem 60 engines; raft subframe system; elastomeric engine mountings; transmission health monitoring

Product:	Kongskilde rotacrat (This unit is made to fit after the harrow behind a tractor; after the harrow pulls up the soil, the rotacraft crumbles it)
Designer:	Finn Ulrich Jensen
Client:	Kongskilde Maskinfabrik A/S Sorø, Denmark
Awards:	1982 Danish Design Council Award
Materials:	Round spring bars for the rotarods; sectional discs stabilized by the counter-pressures of the rotarods; modular construction

Product:	RO-BOOM, Roulund floating oil boom (This product is used to trap and halt the spread of oil slicks emitted by drilling rigs and ships)
Designers:	H. O. Holgersen and L. B. Rasmussen
Client:	A/S Roulunds Fabriker Odense, Denmark
Awards:	1982 Danish Design Council award
Materials:	Rubber; modular construction for easy transportation

Product:	HUP-180 portable hydraulic pump
Designers:	Joe Sonderman and Ed Montague Design/Joe Sonderman, Inc., Charlotte, North Carolina Al Lefler, Fred Schultze, and Darril Plummer Duff-Norton Company, Charlotte, North Carolina
Client:	Duff-Norton Company Charlotte, North Carolina
Materials:	Cover piece: rotationally molded low-intensity linear polyethylene with inherent texture (in mold); reservoir piece: sand-cast aluminum reservoir dipped in Acid F bath to eliminate flash removal abrasions and achieve matte silver finish; black plastic valve control knob; steel toggle power switch with black rubber boot

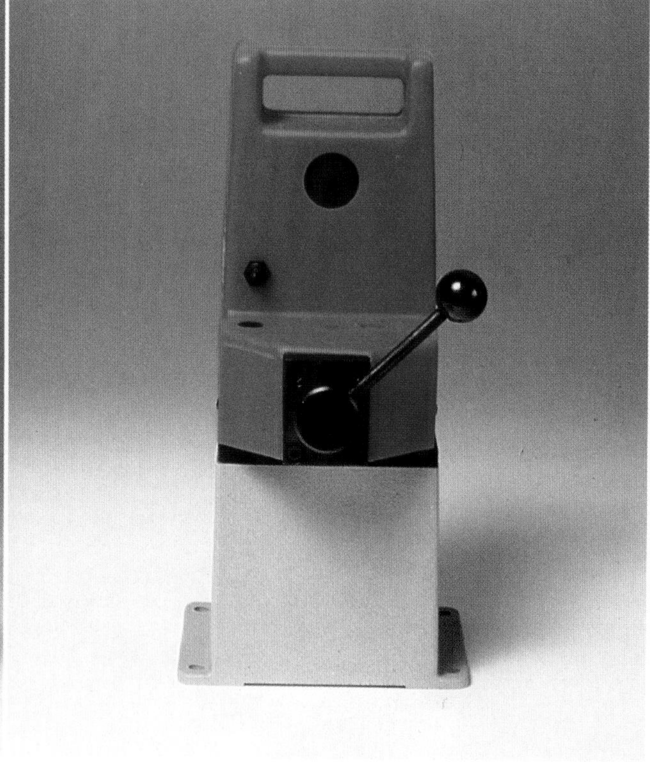

Product: Airless I MK II line marking machine
Designer: Brian Glassel
Client: Dyco Products
Melbourne, Australia
Materials: Steel chassis with fiberglass covers

G4212
LancerBoss
LancerBo

G42
LancerBos

Product:	G. Series frontlift trucks
Designers:	W. Wardle, B. Lowes, P. Ralph
Design Firm:	Peter Ralph Design Associates
Client:	LancerBoss Ltd. Bedfordshire, England
Awards:	1983 Design Council Award
Materials:	Acoustically and thermally insulated cab; driving seat and control console rotate through 180 degrees

Product: Road Tanker
Designers: Paul Holdstock, Michael Aldersley, Brian Field, Maurice Fullwood, Leonard Roe, Robert Small, and Nigel Whal!
Client: M & G Tankers Ltd.
West Midlands, England
Awards: 1983 Design Council Commendation
Materials: Fiber reinforced polyester; double-skinned with the outer shell containing separate seamless inner cells; a semi-rigid polyurethane foam injected around cells binds them to the outer shell

Product: Series TS Turret sideloader
Design Firm: Richardson/Smith
Worthington, Ohio
Client: Crown Controls Corporation
New Bremen, Ohio
Materials: Frame and mast: steel; access panels, seat covers, instrument panels: vacuum-formed ABS plastic

Product:	Articulated 3-axle "Quick Drop" hopper (This system is used for loading, transporting, and discharging ores in underground mines)
Designer:	P. Prins Brakpan, South Africa
Client:	Austral Engineering Works (Pty) Brakpan, South Africa
Awards:	1982 Shell Design Award for Engineering Product
Materials:	Two-module assembly, pivotally connected to allow the system to articulate; to control articulation, a third wheel set and assembly is placed between modules and is pivotally connected to them. The configuration permits more accurate steering and stability on sharp turns and uneven tracks.

Product:	Towing tractor equipment for forestry work, prototype
Designer:	Peter Tucny Prag/CSSR, West Germany
Awards:	1983 Braun Citation for Technical Design

Product: Explosion-proof forklift truck, prototype
Designer: Wolfgang Hesse and Erich Kruse
Braunschweig, West Germany
Awards:: 1983 Braun Prize for Technical Design

Product: Austpole rebutting machine
(This product is used to reclaim power poles with groundline problems caused by rot, accident, or fire)
Client: Austpole Pty. Ltd
Victoria, Australia

CHAPTER 8

Recreational and Sports Equipment

The contemporary passion for physical fitness has spawned not only good health for its practitioners, but also a renewed attention to the gear and equipment with which it can be achieved. Not surprisingly, the exuberance that accompanies the physical activity appears to be matched by a rejuvenation in the design of sports gear.

Designers who recognize the growing marketability of sports products find that using new materials is one way to update old products. Note the Force Fin designed by Bob Evans. Made of polyurethane instead of rubber, it is much more flexible than the conventional fin—and without the conventional ribbing, it can move in any direction, without veering at awkward angles. Moreover, as the designer points out, "The Force Fin folds down on the upkick to reduce water resistance and snaps open for full power on the downstroke. Swimming with the Force Fin at first gives the impression that nothing has been attached to your foot at all, because the water flows backward rather than up and down. You don't have to work hard to make the flexible Force Fin work."

Likewise, the Nike cross country ski racing boot, designed by Trip Allen and Ken Geer, re-examines a traditional piece of equipment. Its redesign is due both to engineering *and* material. As the IDEA jury pointed out, the design addresses "two seemingly incompatible objectives—to simultaneously improve the skier's stability and turning control." The sole of the Nike boot has two parallel ridges running lengthwise along the outer bottom edges. Between these ridges is a cavity that accommodates the ski. By interlocking, the boot and ski become an "integrated turning control system." Moreover, the soles of the boots are unique in their use of Pebax, a specialty resin that has "a feel and performance similar to rubber." Pebax also fuses to itself in molding and eliminates manufacturing problems such as molding-in, or cementing rubber inserts.

The application of new materials, however, is not limited to traditional products. In some cases, such as the "Lance" speed skiing suit designed by Edward Roeanick, new materials are the inspiration for altogether new products. The designer used Dunlopreme 12 foam to form fairings attached to the backs of the arms and legs of the ski suit, and D7 foam in the hood section. A tail was molded in Dunloflex high resilience foam. The result was an aerodynamically shaped suit designed to reduce drag and increase speed, in a sport in which new records are set by milliseconds. Although the suit has not yet contributed to any new records, the idea this prototype promotes is that lightweight, aerodynamically styled foam may do more to enhance speed than the skintight attire favored by most speed skiers. Also implicit is the idea that, by way of new materials, design can suggest the future direction—or further refinement—of certain sports.

Witness as well the Pamir isothermic tunnel tent designed especially to endure—and to let its users endure—extremely cold temperatures. The tent is double-walled, with an outer wall of breatheable silver polyester blizzard that provides thermal insulation, and remains supple even in extremely low temperatures. The inner tent is 100 percent cotton. Two sheets of nylon make up the floor, and for maximum insulation a sheet of closed cellfoam can be inserted between them. In this case, clearly, it is the application of new materials that will permit campers to remain outdoors for longer times and in colder temperatures. Once again, it is through the use of material that the sport allows the user to push his own endurance and skill to even greater limits.

The use of new materials alone does not distinguish contemporary sports gear; physical fitness is achieved these days in new ways and new places, and this, too, determines how the equipment is designed. While deep-sea divers and speed skiers tend to rely upon the local geography to practice their sports and use new high-tech gear, urban dwellers must often make do with the less awesome terrain of an apartment or studio to maintain their physical prowess. The recent array of home exercise tools and machines is evidence that this is indeed possible. Equipment for sports that are to be pursued within these confines, however, must meet a rather different set of demands. That is, these activities do not bring with them the stunning views, breathtaking speeds, or any of the other kinds of gratification offered by outdoor sports. The singular reward is physical fitness, and after a long and demanding day at the office, for example, this reward can pale for even the most disciplined and devoted athlete. So, then, this sports gear must invite and motivate the user, sometimes even suggesting (against all evidence to the contrary) that its operation is effortless. Hence the challenge to the designer. It is one well met by the Amerec 610 Precision Rowing Machine: Not only does the design make its operation self-evident, but the machine also looks comfortable.

Equipment designed for urban athletes must also consider space savings; that it be lightweight, portable, foldable, or otherwise collapsible poses to the designer another set of stringent demands. The foldable bicycle and prototype for a folding motorcycle shown here both offer efficient solutions. Although neither is currently being mass produced, both suggest that the convenience with which such items can be stored may well figure more in the future of their design and marketing.

Overall, the design of contemporary sports equipment appears to be continuing a tradition established in the last 20 years; in which the application of new materials and technologies to the traditional gear introduces untraditional efficiency, speed, strength, or effectiveness, thereby enhancing the skill and technique of the user. Thus, ski poles are bent for aerodynamics; baseball bats, tennis rackets, and skis are no longer necessarily made of wood but of plastics, fiberglass, aluminum, and combinations of unconventional materials. While these certainly add to the durability of the equipment, more to the point is that they add to what the athlete is actually able to achieve—in speed, in strength, or in power. Often, in the end, it is the design process—including the applications of new materials—that most changes the nature of the sport.

Product: Amerec 610 Precision Rowing Machine
Designer: David B. Smith, David Imanaka, and Burns D. Smith
Redmond, Washington
Design Firm: Precor
Redmond, Washington
Client: Amerec Corp.
Bellevue, Washington
Awards: *Industrial Design* magazine
1982 Design Review Selection
Materials: Frame and adjustable clamps: extruded aluminum; rowing arms: stainless steel; bushings, rowing arms, connectors, bumpers: injection-molded Delrin; foot pedals: structural foam; foot straps: velcro; seat: vinyl; elastic shock absorbers: neoprene; seat rollers: injection-molded nylon with capsulated ball bearings

Product: Dunlop Max 200G mid-head graphite tennis racket
Manufacturer: Dunlop Sports Company
West Yorks, England
Materials: Graphite frame filled with polyurethane foam, coated with epoxy paint and synthetic matte lacquer for protection

Product: Dynamics Classics excercise tool
Design Firm: Morison S. Cousins + Associates Inc.
New York, New York
Client: Dynamic Classics
New York, New York
Awards: 1981 *Industrial Design* magazine Design Review selection
Materials: Steel; high-density polyethylene; open cell foam

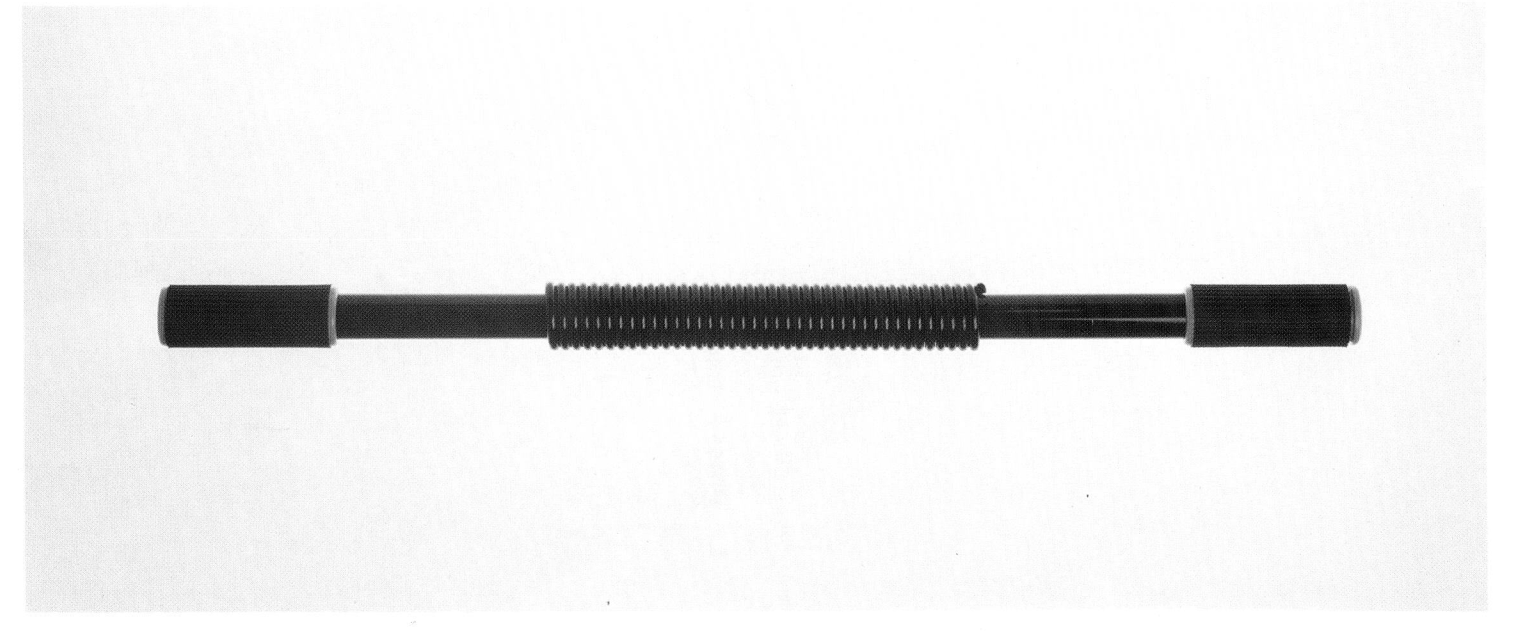

Product: Force Fin
Designer: Bob Evans
Santa Barbara, California
Client: BG Watersports
Redondo Beach, California
Materials: Polyurethane; adjustable nylon straps

Product:	Alpino Pamir III isothermic tunnel tent
Designer:	Andre Heddebaut
Client:	Alpino Renaix, Belgium
Awards:	1983 Belgian Design Centre Signe d'Or
Materials:	Inner tent: 100 percent lightweight cotton; outer tent: silver polyester; floor: two-sheet nylon; frame: lightweight aluminum alloy

Product:	Eclipse Saddlepack
Designer:	W. Shaun Jackson Ann Arbor, Michigan
Design Firm:	Eclipse Inc. Ann Arbor, Michigan
Awards:	1982 IDSA Industrial Design Excellence Award
Materials:	1000 denier Cordura nylon; velcro fasteners; neoprene pad; plated steel hardware

Product:	Nike cross country ski racing boots: Skate (graphite-colored boot) and Kick (white boot)
Designers:	Trip Allen and Ken Geer Exeter, New Hampshire
Client:	Nike, Inc. Exeter, New Hampshire
Awards:	1983 IDSA Industrial Design Excellence Award
Materials:	Synthetic leather, PU coated leather, and silicone treated leather; injection mold finished soles; metal insert in toe piece; thermoplastic PEBAX sole

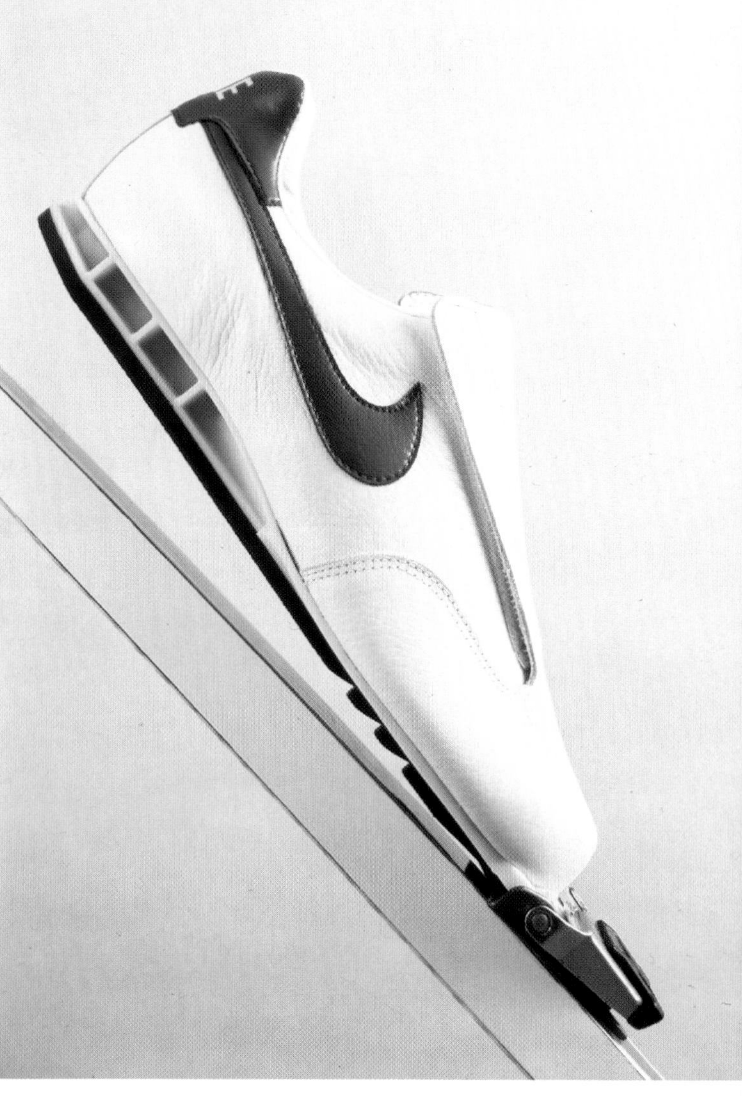

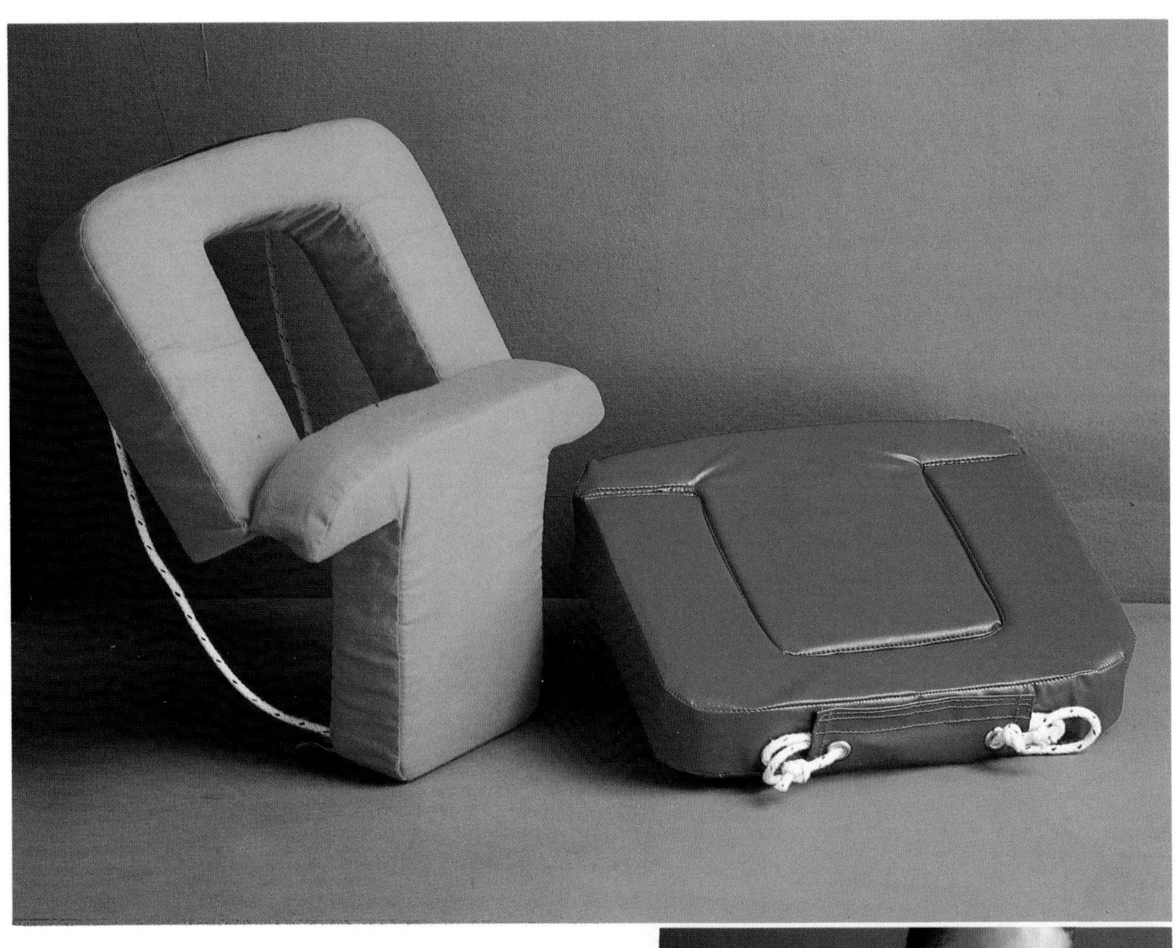

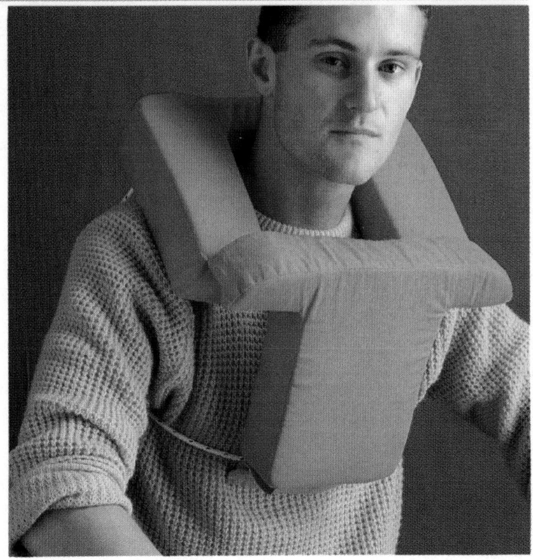

Product:	Safety cushion prototype
Designer:	Francis Carne
Awards:	1983 Dunlopillo Design Award
Materials:	Two interlocking pieces of foam; velcro fasteners

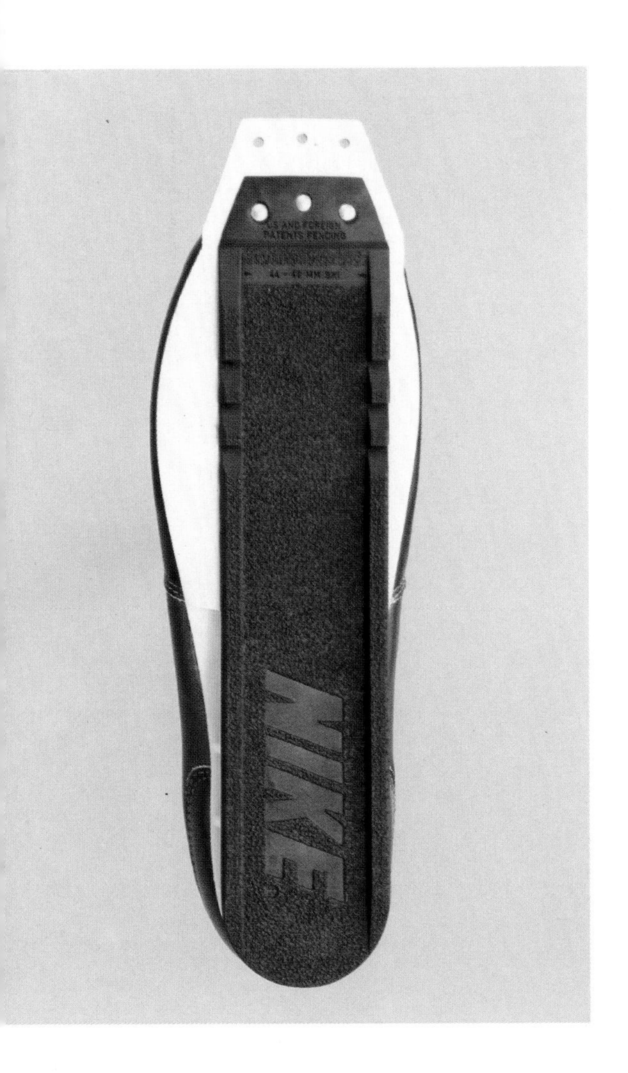

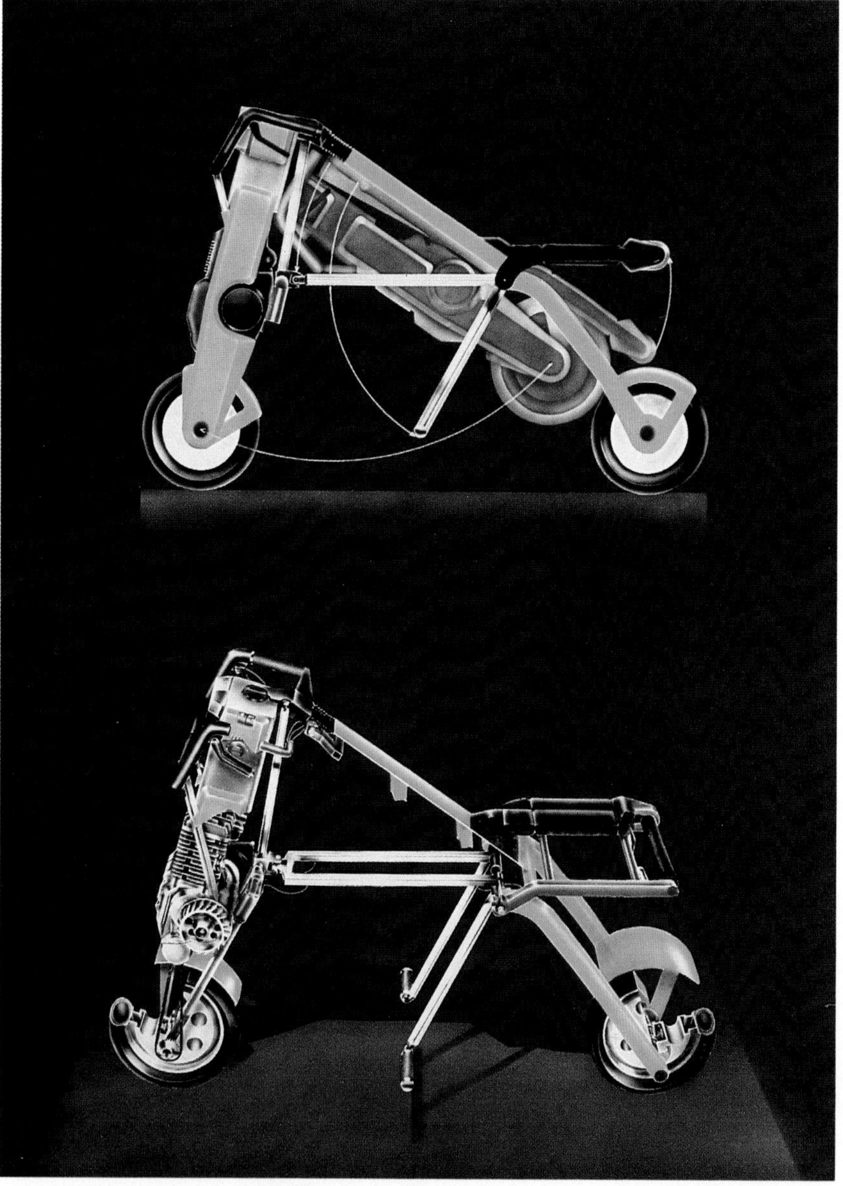

Product:	Foldable urban motorcycle
Designer:	Armando Mercado Villalobos and Antonio Ortiz Certucha Mexico City, Mexico
Design Firm:	Ideograma S.C. Mexico City, Mexico
Awards:	1st International Design Competition award 1982, Osaka, Japan

Product:	Portable bicycle
Designer:	Melvin Best Topanga, California
Design Firm:	Melvin Best Industrial Design Topanga, California
Awards:	1981 *Industrial Design* magazine Design Review selection
Materials:	Die-cast magnesium frame; steel axles, sprockets, bearings; nylon or leather seat and carrying case

Product: Lance ski suit
Designer: Edward Roeanick
Awards: 1983 Dunlopillo Design Award first prize
Materials: Fairings at the backs of arms and legs: Dunlopreme D12 foam; hood section at back of neck: D12 foam; molded Dunloflex high resilience foam; red, satin-finished tricot

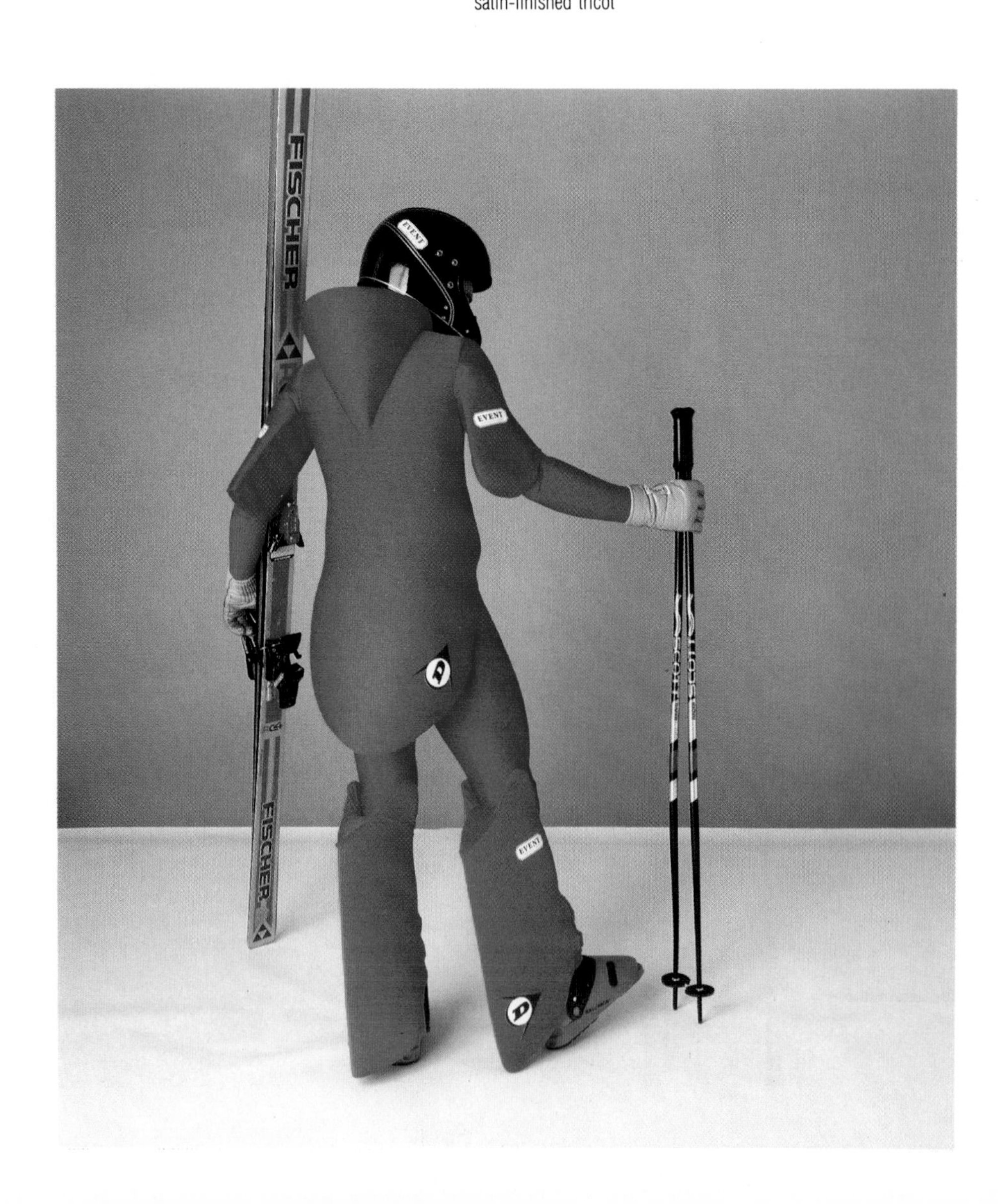

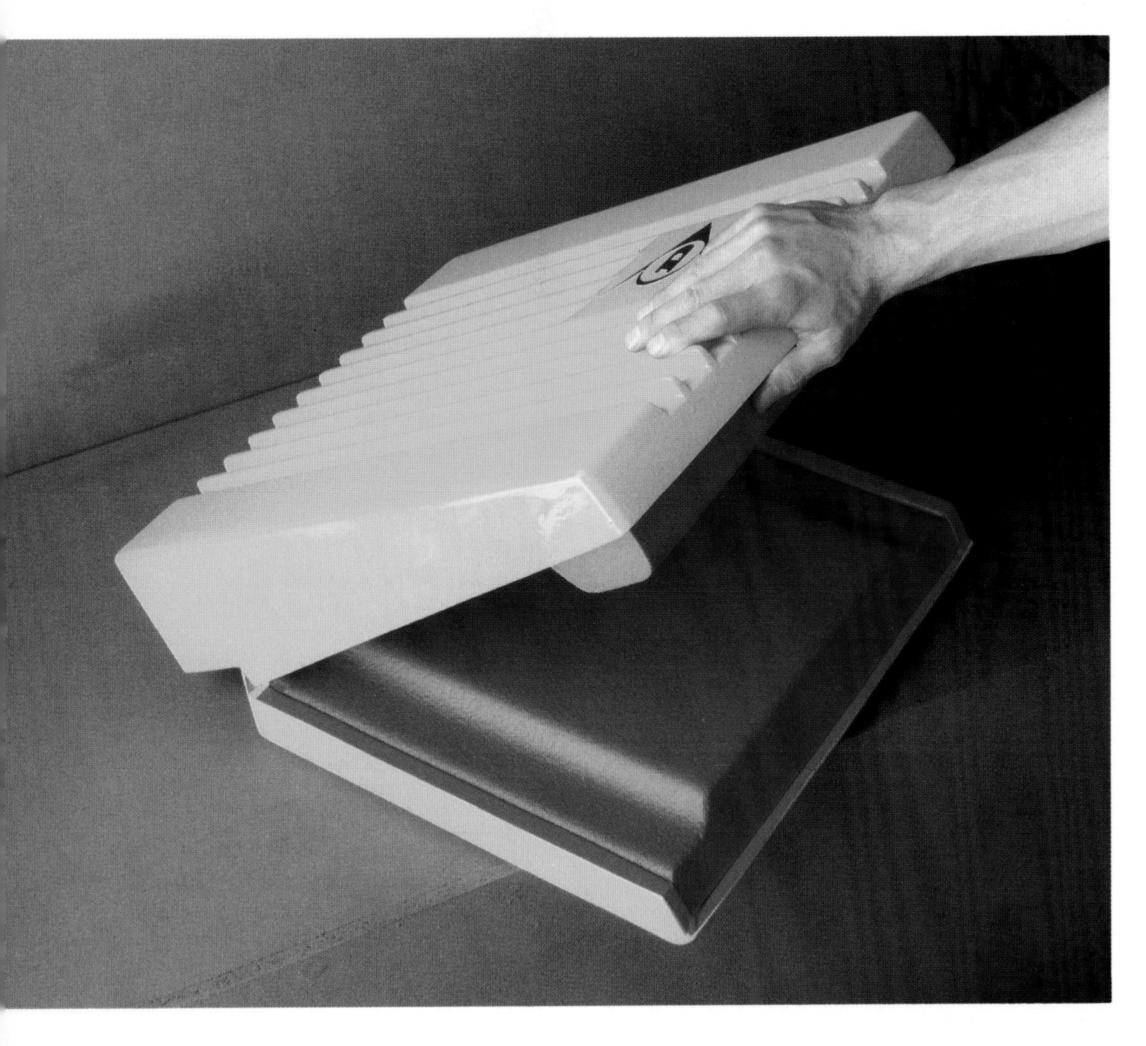

Product:	Stadium seat prototype
Designers:	Dan. St. Zamora, Mircea Arvinte, and Nicolae Barbu
Awards:	1983 Dunlopillo Design Award
Materials:	Injection-molded rigid plastic shell and foam cushions with welded covers; sloping lid allows rain to run off

Product:	*Australia II* keel
Designer:	Ben Lexcen Perth, Australia
Client:	Alan Bond Perth, Australia
Materials:	Aluminum; lead wings

CHAPTER 9

CESHOUSEWARESTOOLSAPPLIANCESHOUSEWARESTOOLSAPPLIANCESHOUSEWARESTOOLSAPPLIANCESHOUSEWAREST
CESHOUSEWARESTOOLSAPPLIANCESHOUSEWARESTOOLSAPPLIANCESHOUSEWARESTOO

CTRONICSENTERTAINMENTHOMEELECTRONICSENTERTAINMENTHOMEELECTRONICSENT
INMENTHOMEELECTRONICSENTERTAINMENTHOMEELECTRONICSENTERTAINMENTHOME

Textiles

GLIGHTINGLIGHTINGLIGHTINGLIGHTINGLIGHTINGLIGHTINGLIGHTINGLIGHTINGLIGHTINGLIGHTINGLIGHTING
GLIGHTINGLIGHTINGLIGHTINGLIGHTINGLIGHTINGLIGHTINGLIGHTINGLIGHTINGLIGHTINGLIGHTINGLIGHTING

TRESIDENTIALFURNISHINGSCONTRACTRESIDENTIALFURNISHINGSCONTRACTRESIDENTIALFURNISHINGSCONTRA
TIALFURNISHINGSCONTRACTRESIDENTIALFURNISHINGSCONTRACTRESIDENTIALFURNISHINGSCONTRACTRESIDE

SEQUIPMENTBUSINESSEQUIPMENTBUSINESSEQUIPMENTBUSINESSEQUIPMENTBUSINESSEQUIPMENTBUSINESSEQU
SEQUIPMENTBUSINESSEQUIPMENTBUSINESSEQUIPMENTBUSINESSEQUIPMENTBUSINESSEQUIPMENTBUSINESSEQU

EQUIPMENTMEDICALEQUIPMENTMEDICALEQUIPMENTMEDICALEQUIPMENTMEDICALEQUIPMENTMEDICALEQUIPMENT
NTMEDICALEQUIPMENTMEDICALEQUIPMENTMEDICALEQUIPMENTMEDICALEQUIPMENTMEDICALEQUIPMENTMEDICAL

IALEQUIPMENTTRANSPORTATIONINDUSTRIALEQUIPMENTTRANSPORTATIONINDUSTRIALEQUIPMENTTRANSPORTAT
IALEQUIPMENTTRANSPORTATIONINDUSTRIALEQUIPMENTTRANSPORTATIONINDUSTRIALEQUIPMENTTRANSPORTAT

IONALSPORTSEQUIPMENTRECREATIONALSPORTSEQUIPMENTRECREATIONALSPORTSEQUIPMENTRECREATIONALSPO
NTRECREATIONALSPORTSEQUIPMENTRECREATIONALSPORTSEQUIPMENTRECREATIONALSPORTSEQUIPMENTRECREA

STEXTILESTEXTILESTEXTILESTEXTILESTEXTILESTEXTILESTEXTILESTEXTILESTEXTILESTEXTILESTEXTILES
STEXTILESTEXTILESTEXTILESTEXTILESTEXTILESTEXTILESTEXTILESTEXTILESTEXTILESTEXTILESTEXTILES

FORTHEHANDICAPPEDDESIGNSFORTHEHANDICAPPEDDESIGNSFORTHEHANDICAPPEDDESIGNSFORTHEHANDICAPPED
ANDICAPPEDDESIGNSFORTHEHANDICAPPEDDESIGNSFORTHEHANDICAPPEDDESIGNSFORTHEHANDICAPPEDDESIGNS

Design of the late 1970s and early 1980s is indelibly marked by a renewed appreciation for color and ornament. Pattern, the decorated surface, and overall visual embellishment have all reappeared in architecture, furnishings, and even business equipment. But there is perhaps no area of design that has benefited more from this new spirit than fabrics and textiles. At long last, the extravagance of the 1960s and the cool colors and beiges of the 1970s have given way to a more exuberant, though controlled, palette.

It is this sense of control, perhaps, that most characterizes the contemporary excitement in color and pattern. As Jack Lenor Larsen and Mildred Constantine point out in their book *The Art Fabric: Mainstream*, "The pendulum has so strongly swung away from the highly charged, romantic expression of the sixties. The new classicism that has replaced it, is cerebral, disciplined, pure, and ordered." It is the antithesis of the more adventuresome, romantic, emotional, and often heroic spirit that preceded it.

Because this appetite for decoration has such a sense of symmetry and suggests a return to pattern, it is only logical that it should make itself so evident in textiles. For textiles—themselves based on the logic of the grid—by nature imply pattern, repetition, and continuity. The simple fact of its weave associates it with logic and order. Still, the sense of pattern and continuity in such evidence in contemporary textiles is not as rigid as it might be. Pattern has reappeared, but it is not as repetitive as one might expect. Subtle shifts in color, gradations in tone, and motifs that are repeated only irregularly are ways in which pattern is established, and then, just slightly, thrown off.

The new appreciation for color in fabrics is especially apparent in contract office applications. That is, upholstery and textiles for the office are now often part of a much larger coordinated color palette. Designer Clino Castelli suggests that it is not only a renewed interest in color, but also an entirely different way of perceiving color. "Color," he says, "is perhaps the single most noticeable imposition of the imposed space. Our reactions to color are emotional, visceral, extremely subjective." He also goes on to attribute this recent change in color preference to the influence of the electronic media. Castelli maintains that the colors of electronic transmission—such as those we see on television—are formed by numerous color overlays, an additive synthesis that is significantly different from "the subtractive colors of the print media." The result is what Castelli calls the "polychromatic tonal tendency," that is, the simultaneous presence of several less saturated colors. Even neutrals, he says, have distinguishable hues. Castelli's program of colors, fabrics, and finishes for the Herman Miller Action Office (shown in Chapter 4) is based on this theory of color. It is a coordination of several tones within one space—a polychromatic rather than a monochromatic environment.

Likewise, the Walker/Group has initiated a Coordinated Design Program with contract manufacturers in carpets, fabrics, laminates, and ceramic tiles. The 21 colors in this palette were selected not only for how they might coordinate with one another; they were also researched and tested for how they might be affected under various contemporary light sources. That is, "pink toned beiges allowed for the effects of fluorescent lighting, clean steel grays, and warm camels without the tendency 'to green.' "

The work of both Castelli and Walker/Group demonstrates how the contemporary use of color in textiles is approached as much as a science as a decorating tool. Larsen, however, also points out how the actual applications of textiles have changed in recent years. "Furnishings must compensate for indoor living in cities by providing organic rhythms, shade and shadow, lively surfaces, fiber itself. Fabrics must maximize light, provide privacy, and enhance a view," he says. "Also, the inherent acoustical and insulating properties of fiber are at last being rediscovered."

What Larsen is pointing out is the growing tendency to use textiles almost as an architectural element. Particularly in cramped urban spaces, visual privacy and separation can often be better achieved by a screen or panel than by constructing a new wall. Acoustical privacy can be accomplished with fabrics as well. Or, as lofts serve as residences more and more in urban areas, their occupants continue to discover how textiles, both as decorative and partitioning devices, are economical and efficient ways to introduce color and texture to expansive interior spaces. Textiles are not merely ornamental, but more and more serve the practical functions traditionally performed by more structural and permanent devices. As Larsen established, "The fabric of floors, walls, windows, and furniture covers will be the means to achieve variation in scale and pattern, color and texture."

The use of materials in textiles appears to be governed by a sense of integrity. That is, while the appeal of natural materials—linens, silks, cottons, and wools—that blossomed in the 1960s remains strong, so too does the appeal of synthetic materials. Yet the latter are used now less to represent another material; synthetic fibers are recognized for their own aesthetics rather than strictly for their imitative capacities. Similarly, contemporary textiles are likely to combine these synthetic fibers with natural fibers in compositions in which each offsets or dramatizes the other. Add this element to the pattern that has reappeared—to the pastel shades of the post modern color palette, to the coordinated color programs found increasingly in interiors, and to the use of fabrics and textiles as architectural devices—and it is evident that this is an area of vigorous and innovative design.

Product: Shibumi
Designer: Developed and adapted by Patricia Green
Client: Groundworks
New York, New York
Awards: 1983 IBD Product Design gold award
Materials: 100 percent cotton

Product: Orion™
Designer: Gary Golkin and Carrie Goldwater Golkin
New York, New York
Client: Art People
New York, New York
Awards: 1983 ASID International Product Design award
Materials: 100 percent cotton

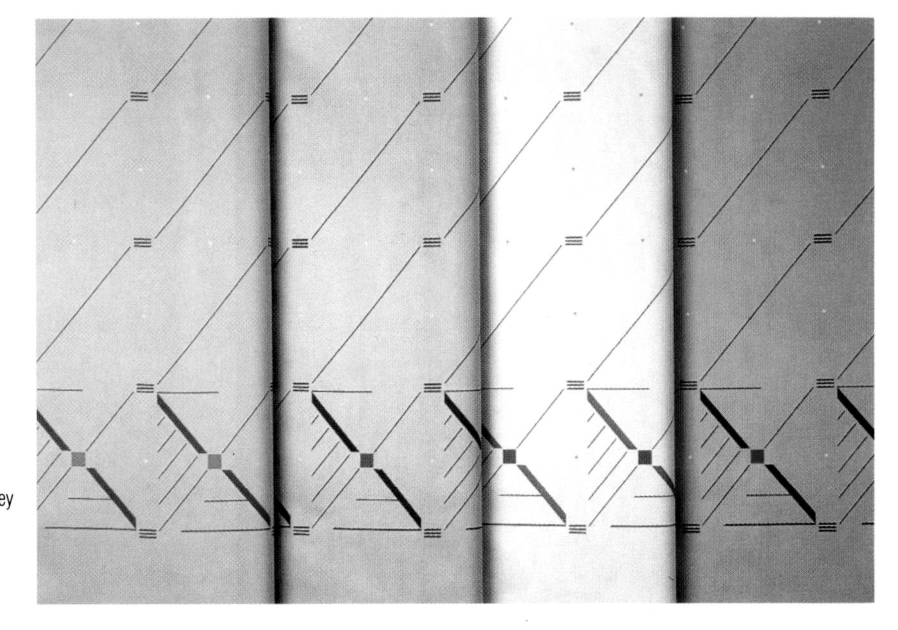

Product: Fret
Designer: Michael Graves
Princeton, New Jersey
Client: Sunar/Hauserman
New York, New York
Materials: 100 percent cotton

Product: Scroll
Designer: Michael Graves
Princeton, New Jersey
Client: Sunar/Hauserman
New York, New York
Materials: 100 percent cotton

Product: Tracery
Designer: Michael Graves
Princeton, New Jersey
Client: Sunar/Hauserman
New York, New York
Materials: 100 percent cotton

Product: Maisema
Designer: Fujiwo Ishimoto
Helsinki, Finland
Client: Marimekko Oy
Helsinki, Finland
Materials: 100 percent cotton, screen-printed

Product: Night Hawk™
Designers: Gary Golkin and Carrie Goldwater Golkin
New York, New York
Client: Art People
New York, New York
Awards: 1983 ASID International Product Design award
Materials: 100 percent cotton canvas

Product: Mira Columba
Designer: Verner Panton
Switzerland
Client: Mira-X
New York, New York
Materials: Printed cotton

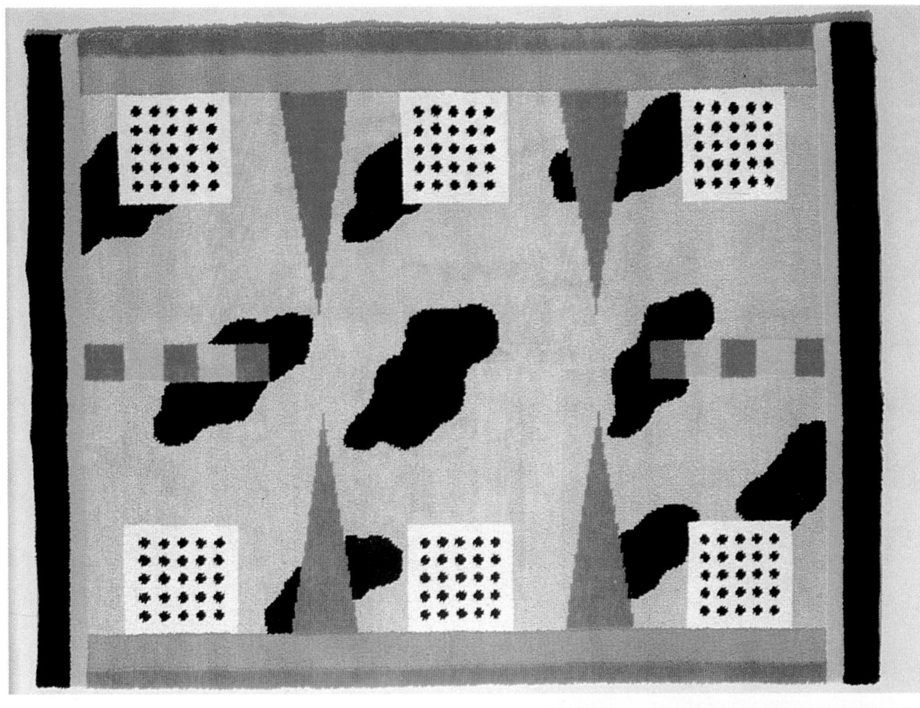

Product: California
Designer: Nathalie du Pasquier
Paris, France
Client: Memphis
Milan, Italy
Materials: Wool carpet, handwoven

Product: Isadora™
Designer: Gretchen Bellinger
New York, New York
Client: Gretchen Bellinger
New York, New York
Awards: 1981 Roscoe Award
Materials: Pleated silk

Product: Valkea Yo
Designer: Fujiwo Ishimoto
Helsinki, Finland
Client: Marimekko Oy
Helsinki, Finland
Awards: 1983 Roscoe Product Design award
Materials: 100 percent cotton

Product: Armor Cloth
Designer: Hazel Siegel
Bedford, New York
Client: DesignTex
Woodside, New York
Awards: 1983 IBD Product Design silver award
Materials: 100 percent nylon with acrylic back

Product: Classic Woolens
Designer: Hazel Siegel
Bedford, New York
Client: DesignTex
Woodside, New York
Awards: 1983 IBD Product Design silver award
Materials: 100 percent wool

Product: Wool Andes
Designer: Hazel Siegel
Bedford, New York
Client: DesignTex
Woodside, New York
Awards: 1983 IBD Product Design honorable mention
Materials: 52 percent wool, 48 percent nylon, with Teflon finish

Product:	Mirage Series
Designer:	Staff design
Client:	DesignTex Woodside, New York
Awards:	1983 IBD Product Design gold award
Materials:	51 percent linen, 49 percent rayon

Product:	Kinnasand Collection
Designer:	Staff design
Client:	DesignTex Woodside, New York
Awards:	1982 ASID International Product Design award 1982 IBD Product Design honorable mention
Materials:	100 percent cotton

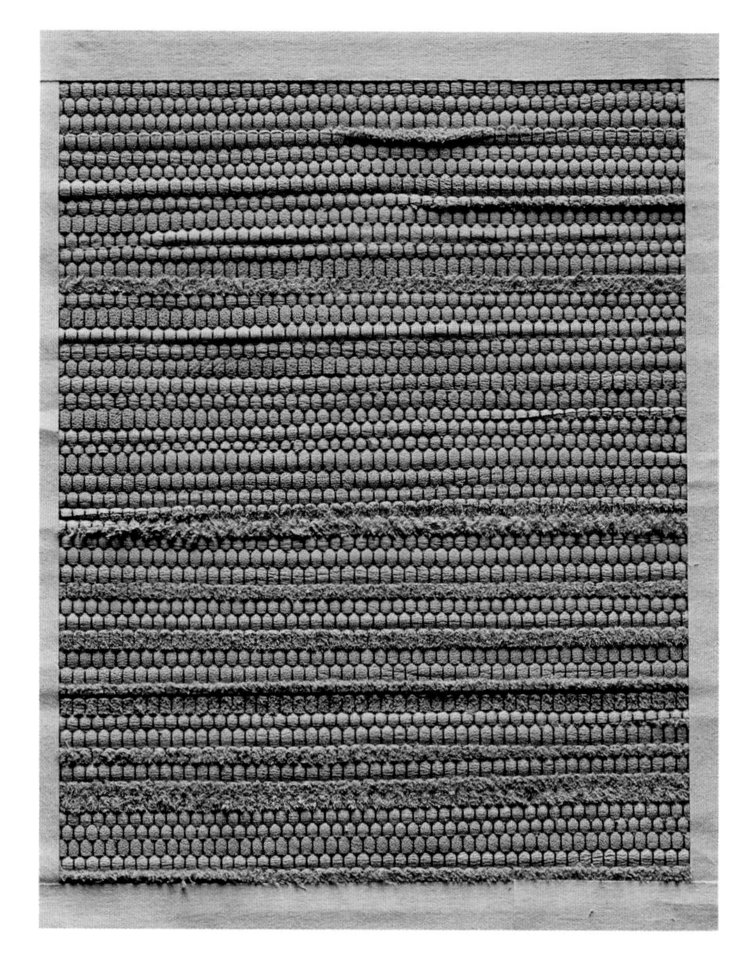

Product:	Wholly Cow
Designer:	Charles Schambourg Brussels, Belgium
Client:	Jack Lenor Larsen New York, New York
Awards:	1982 Roscoe Award 1982 ASID International Product Design award
Materials:	Woven leather

Product:	Stripes
Designers:	Roger McDonald and Doug V'Soske New York, New York
Client:	V'Soske New York, New York
Awards:	1983 IBD Product Design gold award 1983 Roscoe Product Design award
Materials:	Machine tufted custom wool carpet

Products: (L to R) Sundown II, agate; Diamond Jubilee, silverweed; County Downs, straw; Limerick, meadow; Ribcord, moonstone; Merrion Square, silverweed; County Check, straw (front and back sides).
Design Firm: Jack Lenor Larsen design studios
New York, New York
Client: Jack Lenor Larsen
New York, New York
Materials: Sundown II, 100 percent wool; Diamond Jubilee, 100 percent wool; County Downs, 56 percent wool and 44 percent cotton chenille; Limerick, 100 percent wool; Ribcord, 50 percent worsted wool and 50 percent cotton; Merrion square, 100 percent worsted wool; County Check, 57 percent wool and 43 percent cotton chenille

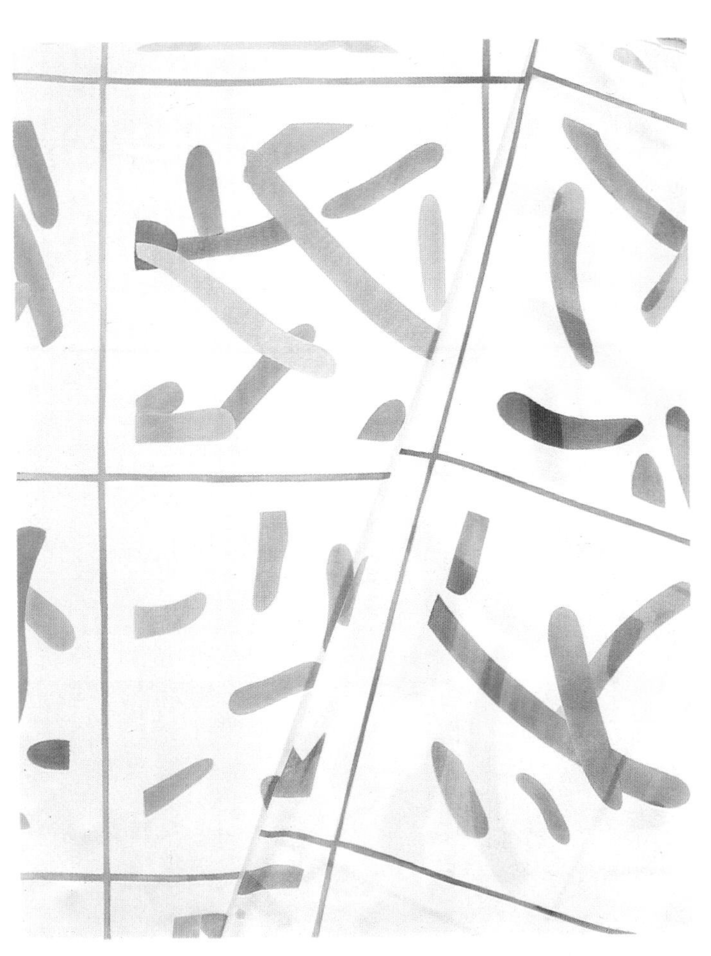

Product: Silhouette
Designer: Staff design
Client: DesignTex
Woodside, New York
Awards: 1982 IBD Product Design silver award
Materials: 54 percent cotton, 46 percent polyester

Product: Coordinated Resources Program
Design Firm: The Walker/Group
New York, New York
Materials: Fabrics, laminates, carpets, and ceramic tile for coordinated interior program.

Product: Stellar Collection: Clockwise: Northern Lights
Chroma, Nova, Hue
Design Firm: Boris Kroll Design Studio
Client: Boris Kroll Fabrics
New York, New York
Materials: Wool blends

Product: Coordinated Resources Program
Design Firm: The Walker/Group
New York, New York
Materials: Fabrics, laminates, carpets, and ceramic tile for coordinated interior programs

Product: Coordinated Resources Program
Design Firm: The Walker/Group
New York, New York
Materials: Fabrics, laminates, carpets, and ceramic tile for coordinated interior programs

Product: Metro
Design Firm: Yves Gonnet Inc.
New York, New York
Client: Yves Gonnet Inc.
New York, New York
Materials: Silk Jacquard taffeta

Product: Coordinated Resources Program
Design Firm: The Walker/Group
New York, New York
Materials: Fabrics, laminates, carpets, and ceramic tile for coordinated interior programs

Product: Coordinated Resources Program
Design Firm: The Walker/Group
New York, New York
Materials: Fabrics, laminates, carpets, and ceramic tile for coordinated interior programs

Product: Prestwick Cloth
Designer: Ward Bennett
New York, New York
Client: Ward Bennett Designs for Brickel Associates
New York, New York
Materials: 70 percent wool, 30 percent nylon (acrylic backing)

Product: Unique Cloth
Designer: Ward Bennett
New York, New York
Client: Ward Bennett Designs for Brickel Associates
New York, New York
Materials: 100 percent cotton chenille

Product: Tapestry Cloth
Designer: Ward Bennett
New York, New York
Client: Ward Bennett Designs for Brickel Associates
New York, New York
Materials: 90 percent wool, 10 percent nylon
Jacquard weave

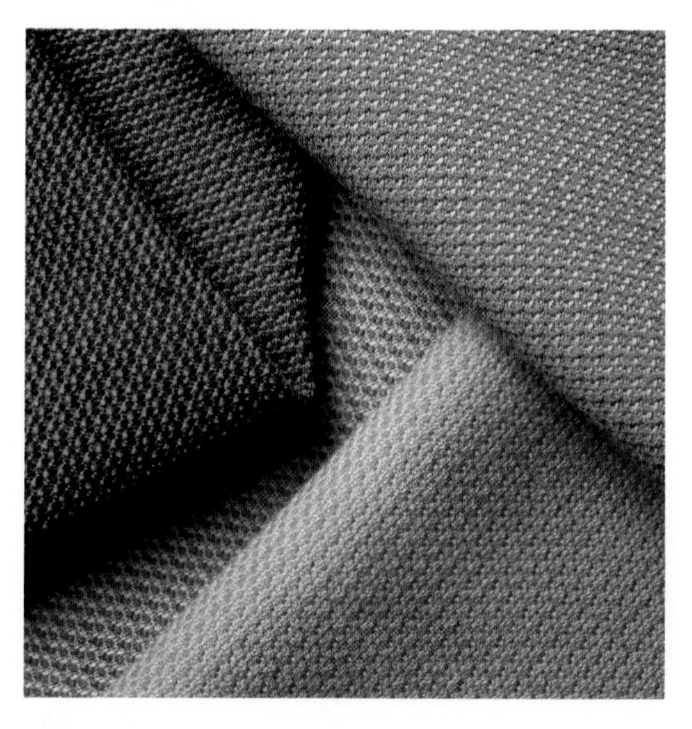

Product: Waffle Weave
Designer: Ward Bennett
Client: Ward Bennett Designs for Brickel Associates
New York, New York
Materials: 100 percent English worsted wool

Product:	Monochrome Range of Curtaining
Designer:	L. Davison, C. Kuhn Mowbray, South Africa
Client:	Sun Fabrics Mowbray, South Africa
Awards:	1982 Shell Consumer Product Design Award
Materials:	100 percent cotton, hand printed

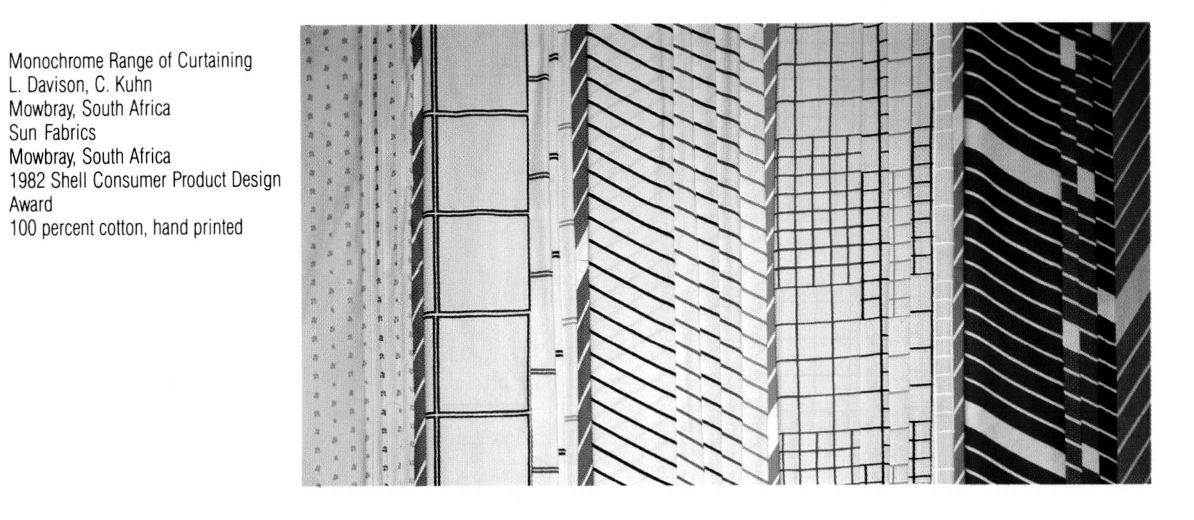

Product:	Jhane Barnes Collection
Designer:	Jhane Barnes New York, New York
Client:	Knoll International New York, New York
Awards:	1983 ASID International Product Design award
Materials:	Lightweight cotton, wool, and rayon

Product:	Diamond Dancer
Design Firm:	Day & Ernst for Patterson, Flynn & Martin New York, New York
Client:	Patterson, Flynn & Martin New York, New York
Awards:	1983 Roscoe Product Design award
Materials:	Handpainted sisal rug

CHAPTER **10**

Designs for the Handicapped

ICESHOUSEWARESTOOLSAPPLIANCESHOUSEWARESTOOLSAPPLIANCE[illegible]USEWAREST[illegible]OLSAPPLIANCESH[illegible]SEWAREST
ICESHOUSEWARESTOOLSAPPLIANCESHOUSEWARESTOOLSAPPLIANCE[illegible]ESH[illegible]T

ECTRONICSENTERTAINMENTHOMEELECTRONICSENTERTAINMENTHOM[illegible]THO[illegible]RON
AINMENTHOMEELECTRONICSENTERTAINMENTHOMEELECTRONICSENT[illegible]SEN[illegible]MEN

IGLIGHTINGLIGHTINGLIGHTINGLIGHTINGLIGHTINGLIGHTINGLIGHTINGLIGHTINGLIGHTINGLIGHTINGLIGHTINGLIGHTING
IGLIGHTINGLIGHTINGLIGHTINGLIGHTINGLIGH[illegible]G

CTRESIDENTIALFURNISHINGSCONTRACTRESIDE[illegible]SH[illegible]A
ITIALFURNISHINGSCONTRACTRESIDENTIALFUR[illegible]ONT[illegible]E

SSEQUIPMENTBUSINESSEQUIPMENTBUSINESSEQUIPMENTBUSINESSEQUIPMENTBUSINESSEQUIPMENTBUSINESSEQU
SSEQUIPMENTBUSINESSEQUIPMENTBUSINESSEQUIPMENTBUSINESSEQUIPMENTBUSINESSEQUIPMENTBUSINESSEQU

EQUIPMENTMEDICALEQUIPMENTMEDICALEQUIPMENTMEDICALEQUIPMENTMEDICALEQUIPMENTMEDICALEQUIPMENT
ENTMEDICALEQUIPMENTMEDICALEQUIPMENTMEDICALEQUIPMENTMEDICALEQUIPMENTMEDICALEQUIPMENTMEDICAL

RIALEQUIPMENTTRANSPORTATIONINDUSTRIALEQUIPMENTTRANSPORTATIONINDUSTRIALEQUIPMENTTRANSPORTAT
RIALEQUIPMENTTRANSPORTATIONINDUSTRIALEQUIPMENTTRANSPORTATIONINDUSTRIALEQUIPMENTTRANSPORTAT

TIONALSPORTSEQUIPMENTRECREATIONALSPORTSEQUIPMENTRECREATIONALSPORTSEQUIPMENTRECREATIONALSPO
ENTRECREATIONALSPORTSEQUIPMENTRECREATIONALSPORTSEQUIPMENTRECREATIONALSPORTSEQUIPMENTRECREA

ESTEXTILESTEXTILESTEXTILESTEXTILESTEXTILESTEXTILESTEXTILESTEXTILESTEXTILESTEXTILESTEXTILESTEXTILES
ESTEXTILESTEXTILESTEXTILESTEXTILESTEXTILESTEXTILESTEXTILESTEXTILESTEXTILESTEXTILESTEXTILESTEXTILES

SFORTHEHANDICAPPEDDESIGNSFORTHEHANDICAPPEDDESIGNSFORTHEHANDICAPPEDDESIGNSFORTHEHANDICAPPED
HANDICAPPEDDESIGNSFORTHEHANDICAPPEDDESIGNSFORTHEHANDICAPPEDDESIGNSFORTHEHANDICAPPEDDESIGNS

In the discussion of contemporary product design, what is perhaps most compelling about design for the handicapped is that it has finally become a distinct and separate category in itself. Until recently, design for the handicapped referred largely to furniture that had been retrofitted for the impaired, or equipment whose highly clinical appearance, although it might accommodate a physical handicap, did little to psychologically accommodate the user. This is no longer the case. Take, for example, the walker designed for children made by Habermaass of West Germany. Until recently, most walkers had been aluminum. This one, however, is wood and has been painted bright red. It also has an attachment for a toy wagon. These new amenities help to remove the clinical stigma from the product, make it less intimidating to the child, and encourage its use.

Perhaps one reason that design for the handicapped has remained such uncharted territory is that standard anthropometrics rarely apply to the infirm. Designers who address the handicapped must frequently conduct their own research. In designing bathing fixtures for the handicapped, for example, Pascal Malassigne and James Bostrum discovered that, generally speaking, the size of people in the handicapped group they worked with was smaller than able-bodied subjects. The grants and financial support necessary to conduct the research that yields this information is not often so readily available to designers. That research information was available in this case, and that the new information was reflected in the design of the fixtures is not only encouraging; it also suggests that careful human factors research is likely to become and remain an integral part of designing for the handicapped.

Yet despite the fact that designers and critics have recognized design for the handicapped as an area that pleads for attention, products designed especially for the handicapped are not the only answers to the problem. More often, products designed for the able-bodied can accommodate the handicapped user as well. This, perhaps, is even more reassuring than the whole range of new products for the handicapped, simply because it demonstrates a more human approach. That is, disabilities can be accommodated by good design; they do not necessarily signal an abnormal condition that demands unique measures. The use of color, for example, is one simple way to answer certain needs of the elderly. The fading eyesight which often accompanies old age tends to make it difficult to distinguish the cooler colors, and diminishes perception of color intensity. Many colors look gray. Studies have indicated, not surprisingly, that the elderly often prefer more intense colors than do young people. Therefore, the increased use of bright colors in product and package design, while answering to a more general contemporary appetite for color and ornamentation, might also answer to this specific need of the elderly.

Victor Papanek states in his book *Design for Human Scale*, "Design is to technology what ecology is to biology." Nowhere does this become more evident than in design for the handicapped. To begin with, the term "handicapped" has a loose definition; it includes not only those with permanent disabilities, but can refer to the temporarily infirm, pregnant women, people under five feet tall perhaps, and quite often the elderly, who represent an increasingly important market as the population grows older. In other words, almost all of us, at some point or another in our lives, may be "handicapped." Although this may seem obvious, it has only recently been recognized and addressed as such.

Contemporary design for the handicapped also recognizes, finally, ergonomics. Because products and furnishings for the handicapped may be the areas that most need ergonomic design, it is ironic that they have been so long without it. As Bill Bash, designer of the Titann wheelchair, points out, most wheelchairs "though designed to sit in all day, are no more comfortable than your average church pew." The Titann wheelchair that he subsequently designed considers weight, portability, comfort, and durability. It is also narrower than standard wheelchairs and is foldable, permitting more convenient travel—and thus independence—for its user. Finally, the center of gravity on the chair is adjustable, allowing people of different height, build, and body structure to be comfortable in it.

Designers at Douglass Ball, Inc., in Canada have also addressed the transportation problem of wheelchair users. Their wheelchair restraint system designed for trains, buses, and specialized vehicles can adjust to a different number of wheelchair models. Telescoping arms fasten and lock the wheelchair into position against the wall, and a three-point belt system secures the passenger in his seat. Likewise, an elevator system designed by Douglass Ball recognizes "the fear, feelings of insecurity, frustration, and helplessness" of wheelchair users and the semiambulatory when boarding a bus, and was designed to allow passengers "to board in a dignified manner with movements of the elevator they are familiar with."

As all of these products demonstrate, user independence is the first objective in many of these designs. And whether they allow handicapped persons to drive, bathe, walk, or wheel themselves about more easily and comfortably, they are products that not only accommodate disabilities, but permit and encourage user independence and autonomy despite them. This shift in emphasis, too, distinguishes contemporary design for the handicapped.

Occasionally, and the occasions are certainly atypical, products designed for the handicapped, though intending only to afford greater comfort and independence, achieve even more substantial results. The ORLAU Swivel Walker is a case in point. The walker is designed to enable patients who are totally paralyzed from the waist down to walk. Although the walker can certainly be used by full-grown adults, its rewards are greatest when it is used by children at the age of 15 months and upward. As the walker's client—an English orthopedic hospital—maintains, the walker "provides ambulation which matches the normal development pattern. It is also beneficial in putting mechanical stress through the skeleton, thus substantially eliminating so-called spontaneous fractures which have so bedevilled ambulatory progress. And, the ability to be upright at an early age improves drainage of the urinary system, failure of which is a considerable potential danger." As great as (if not greater than) the comfort and independence allowed by the walker are its actual therapeutic advantages. The fact that products designed for the handicapped can have such therapeutic value places greater challenges—and gives greater rewards—to the industrial designer than those he might encounter designing more conventional products and furnishings.

Although it is not the case with the ORLAU Walker, a number of the products shown in this chapter are prototypes, rather than manufactured products which can be found on the market. They are included here, nevertheless, because design for the handicapped is an area so ripe for innovation and development that even prototypes, which may do nothing more than suggest new directions, are as important as those products already on the market.

Product:	EVAC + Chair
Designer:	David Egen New York, New York
Design Firm:	Egen Polymatic Corporation New York, New York
Awards:	1982 IDSA Industrial Design Excellence Award 1982 *Time* magazine selection for "The Best of 1982"
Materials:	Aluminum tubing; seat: vinyl covered nylon pocket

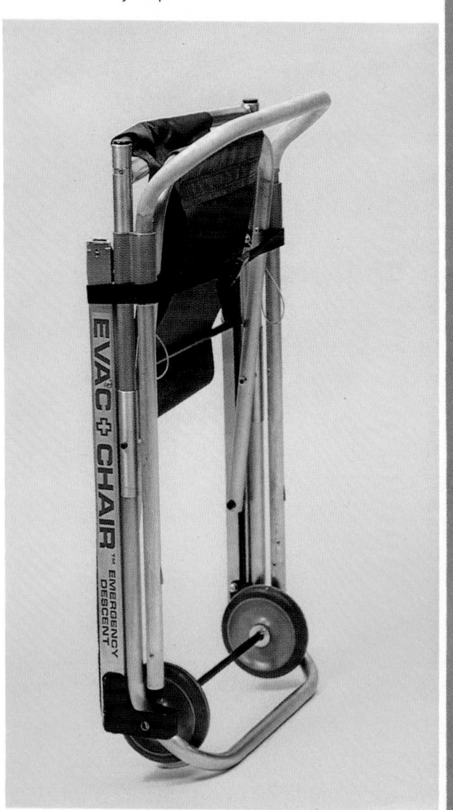

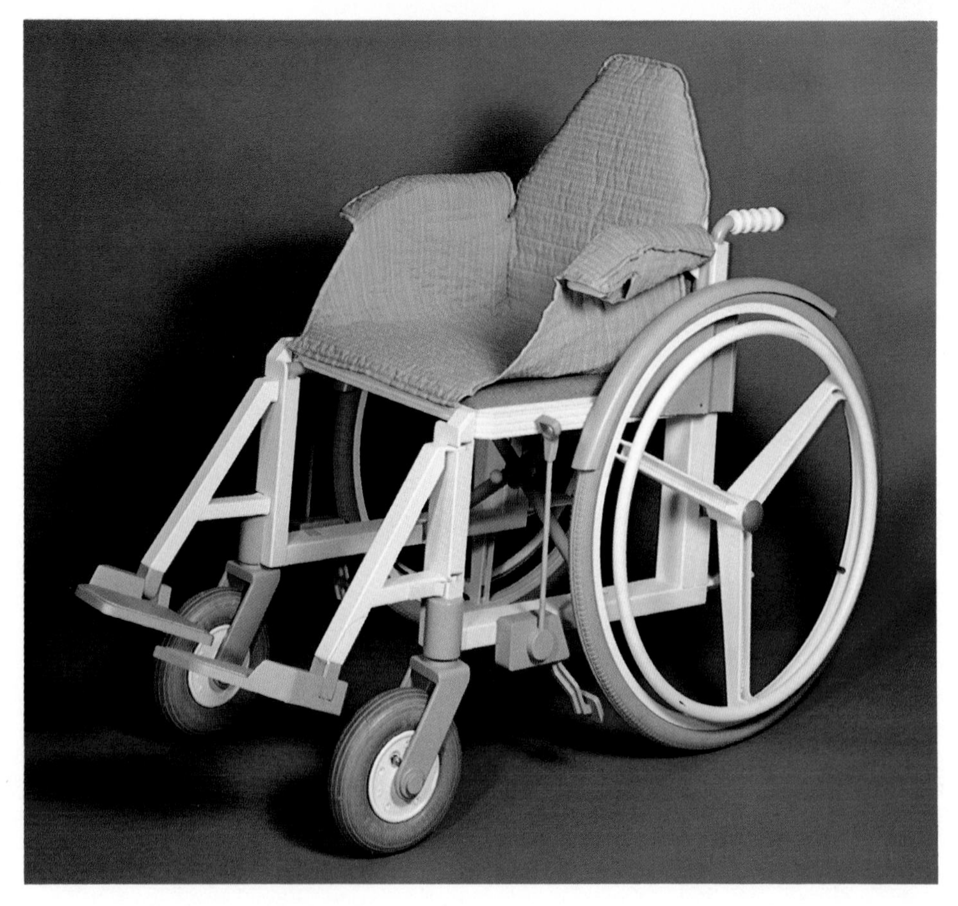

Product:	Foldable standard wheelchair
Designer:	Jorg Ratzlaff and Frank Rieser Hamburg, Federal Republic of Germany
Awards:	1983 Braun Prize for Technical Design

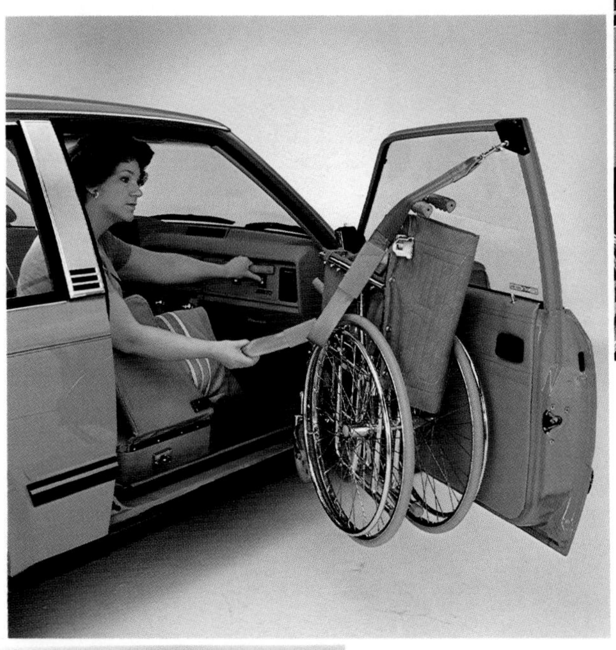

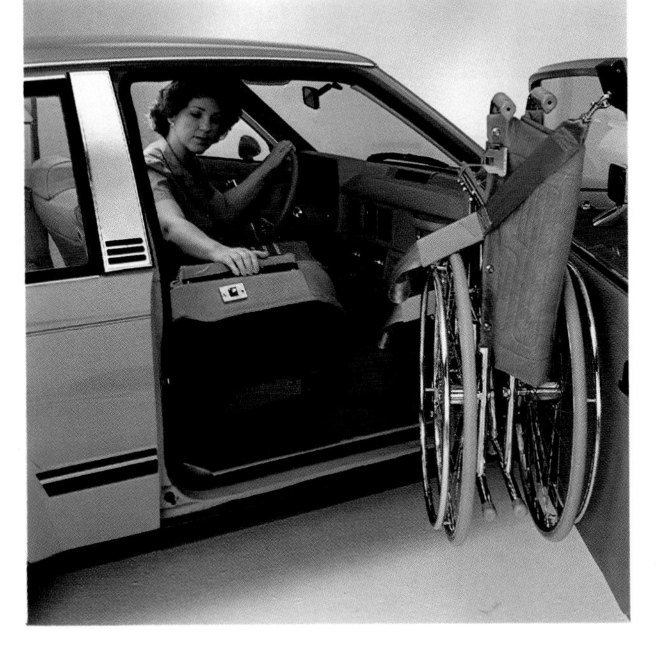

Product: IMP II Improved Mobility Package (This is for handicapped drivers for docking, transfer and retrieval, and storage.)
Designers: Joint design development between The Ford Motor Company Dearborn, Michigan and ASC Inc. Southgate, Michigan
Manufactured and installed by: Wisco Corporation Subsidiary of ASC Inc. Ferndale, Michigan
Materials: Power adjustable bench seat with integral padded transfer bridge; power wheelchair retriever and retaining system

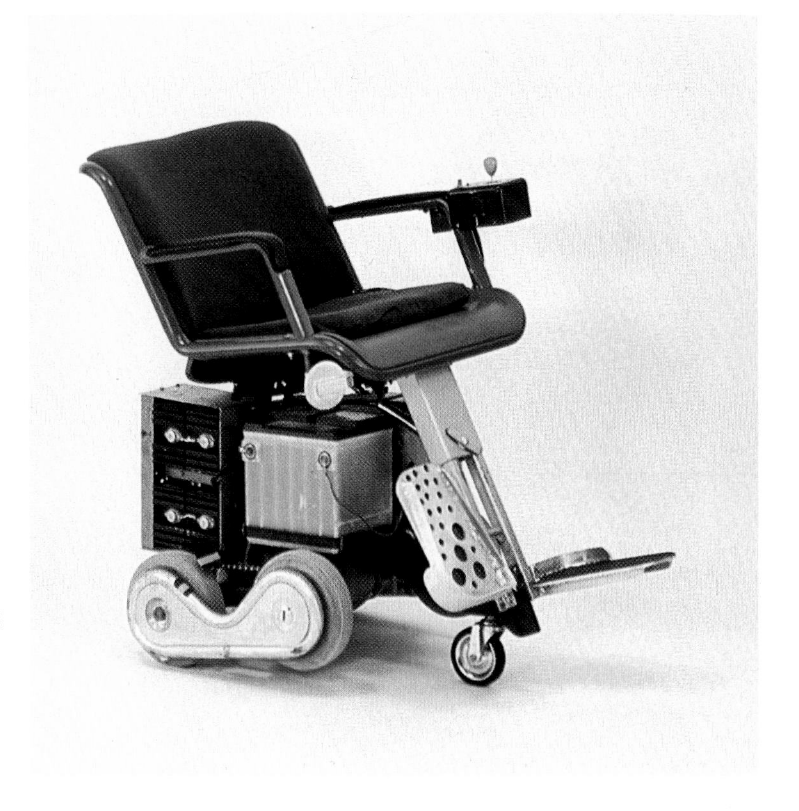

Product: Electric curb-climbing wheelchair (This chair is equipped with small front wheel for indoor use.)
Design Firm: Douglass Ball Inc. Sainte-Anne-de-Bellevue, Quebec, Canada
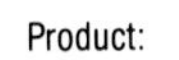
Materials: Lightweight aluminum and steel

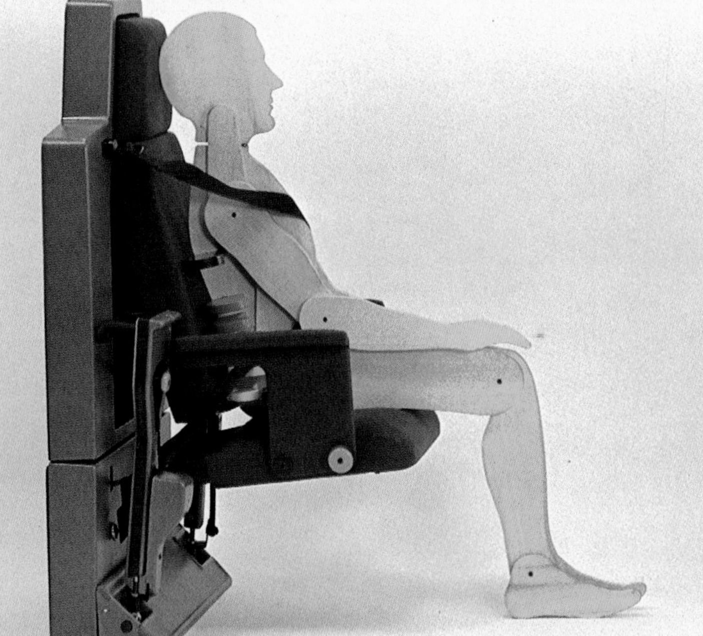

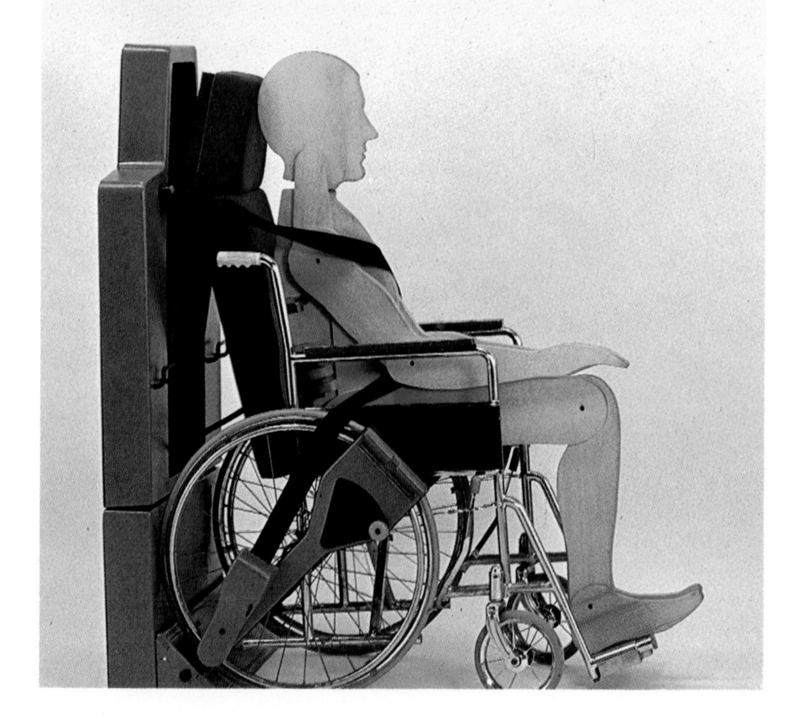

Product:	Wheelchair restraint system for trains and buses (This chair is equipped with adjustable head and backrest, three-point belt, and telescoping arms. The system is also designed for side-by-side installation and can also be converted into a special seat with pivoting armrest for semi-ambulatory passengers.)
Design Firm:	Douglass Ball Inc. Sainte-Anne-de-Bellevue, Quebec, Canada
Materials:	Steel frame covered with ABS plastic

Product:	Child's walker
Designer:	Horst Dwinger Rodach, West Germany
Client:	Habermaass Rodach, West Germany
Courtesy:	The Able Child New York, New York
Materials:	Wood

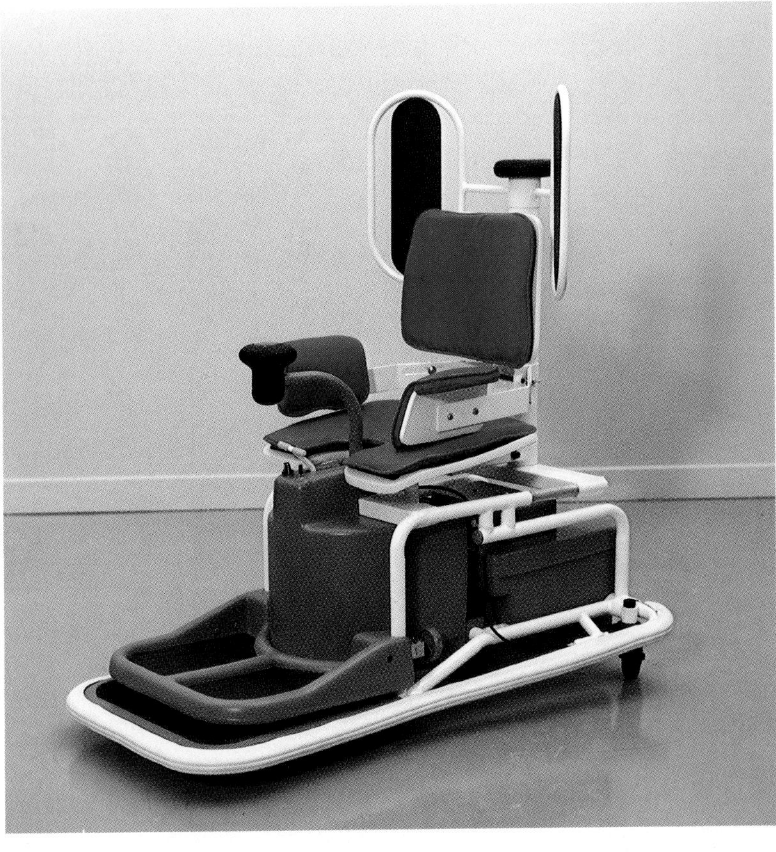

Product: Chid's wheelchair
(This chair comes with adjustable seat and back and armrest supports.)
Design Firm: Douglass Ball Inc.
Sainte-Anne-de-Bellevue, Quebec, Canada
Materials: Molded foam; fabric; steel tubing

Product: Titann Wheelchair
Designer: Bill Bash
Lakeville, Minnesota
Client: Theradyne Corporation
Lakeville, Minnesota
Materials: Black urethane spherical castor; titanium frame

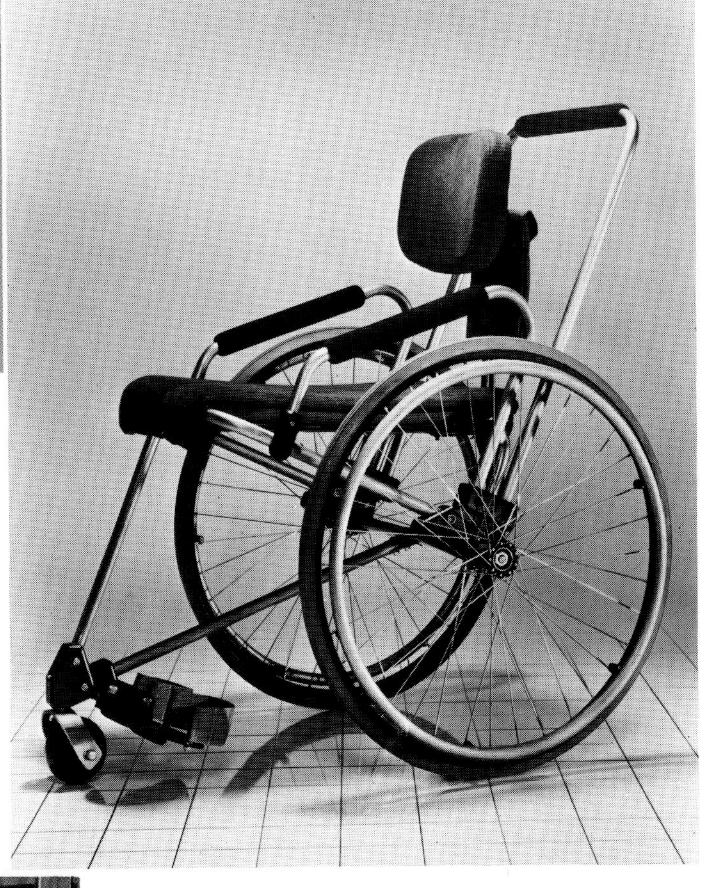

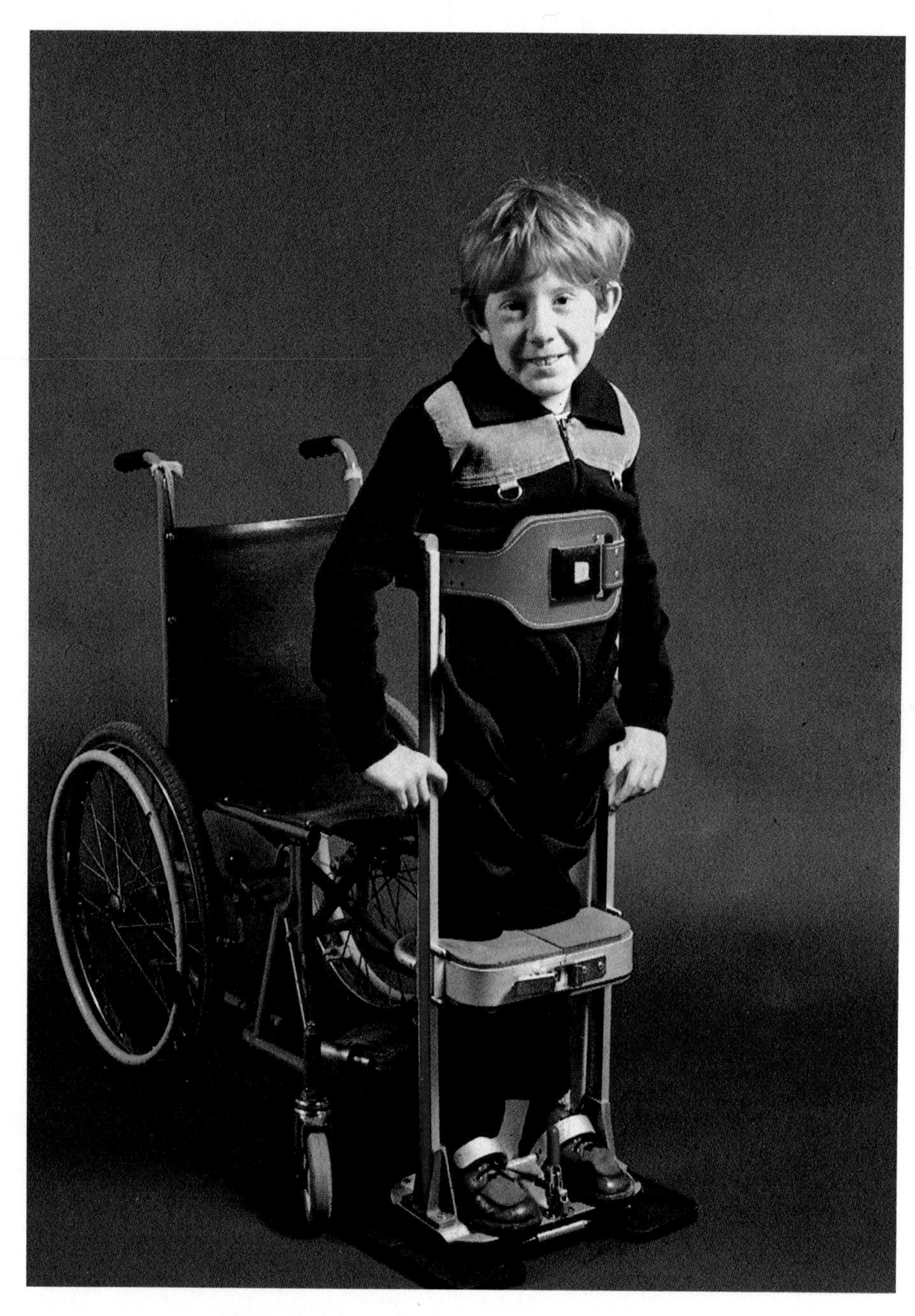

Product: ORLAU Swivel Walker
Designer: G. K. Rose and Engineering Team
Client: Orlau, The Orthopedic Hospital
Owestry, Shropshire, England
Awards: 1981 Design Council award
Materials: Steel and aluminum

Product: Roll-In Shower
(This is made for disabled people unable to transfer safely from a wheelchair.)
Designers: James A. Bostrum and
Pascal Malassigne
Atlanta, Georgia
Project Engineer: William Sevebeck
Blacksburg, Virginia
Materials: Fiberglass shells; wood floor and ramp
Grants from the National Institute of Handicapped Research, U.S. Department of Education, and the Office of Rehabilitation Research and Development, the Veterans Administration, made this project possible.

Product: Transfer Tub Seat
Designers: Ronald L. Mace and Nancy Hurley
Raleigh, North Carolina
Design Firm: Barrier Free Environments, Inc.
Raleigh, North Carolina
Awards: 1983 *Industrial Design* magazine
Design Review selection
Materials: Seat and transfer board: cross-linked polyurethane; adjustable base and clamps: stainless steel; armrest: stainless steel tube padded with vinyl-coated foam pad

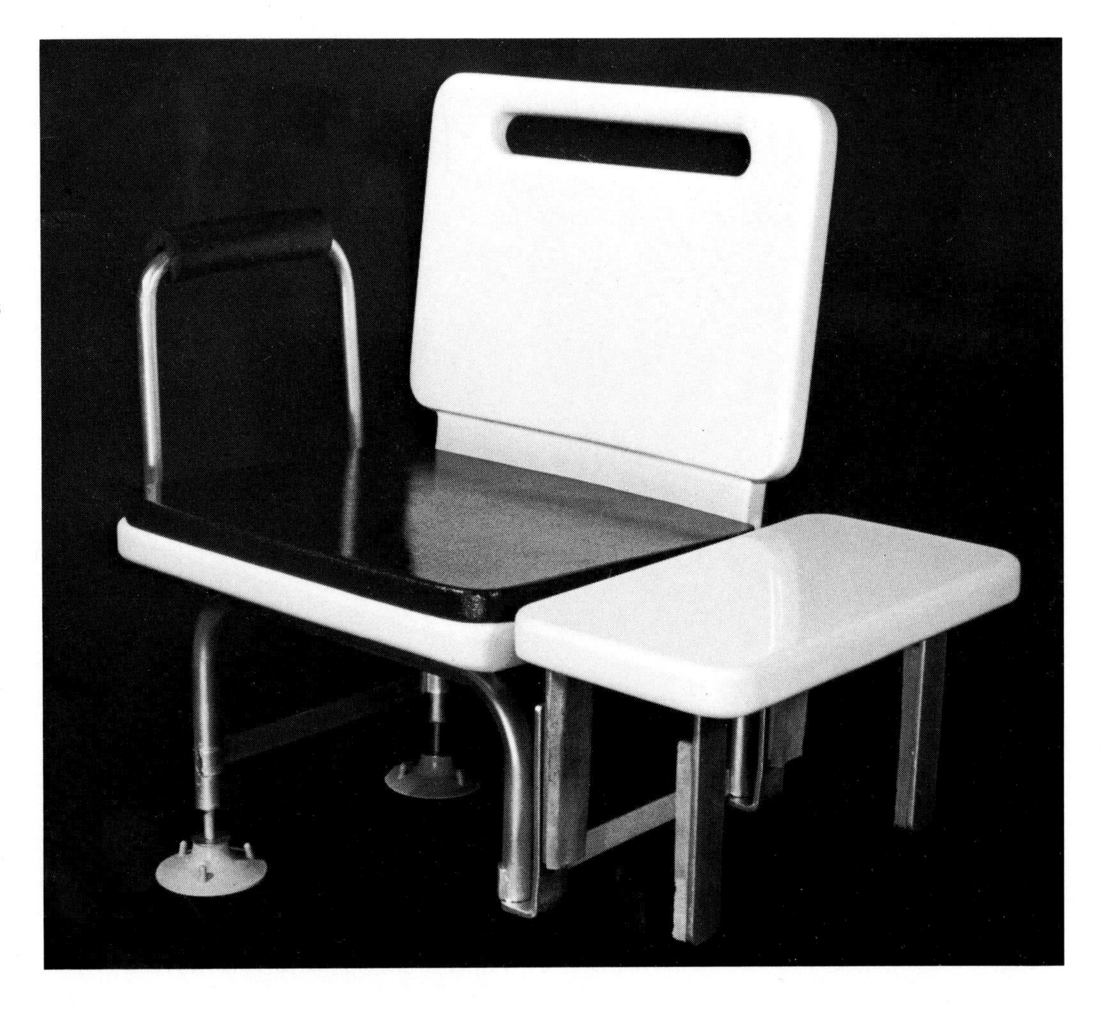

Product: Co-driver vehicle control device
(This is designed to allow persons with lower-body and lower-limb disabilities to drive an automatic transmission vehicle using hands only.)
Designer: N. Cohen
Linksfield, South Africa
Awards: 1982 Shell Design Consumer Product award
Materials: Steel and aluminum

Product: Polyart abstract pattern jigsaw
(These are used to encourage basic manipulation as well as perceptual-motor skills)
Materials: Plywood
Courtesy The Able Child, New York

Product: Seat for shower stall prototype
(This is designed for installation in an existing shower stall for use by semi- and nonambulatory individuals who can transfer from a wheelchair.
Designers: James A. Bostrum and
Pascal Malassigne
Atlanta, Georgia
Project Engineer: William Sevebeck
Blacksburg, Virginia
Materials: Fiberglass
Grants from the National Institute of Handicapped Research, U.S. Department of Education, and the Office of Rehabilitation Research and Development, the Veterans Administration, made this project possible.

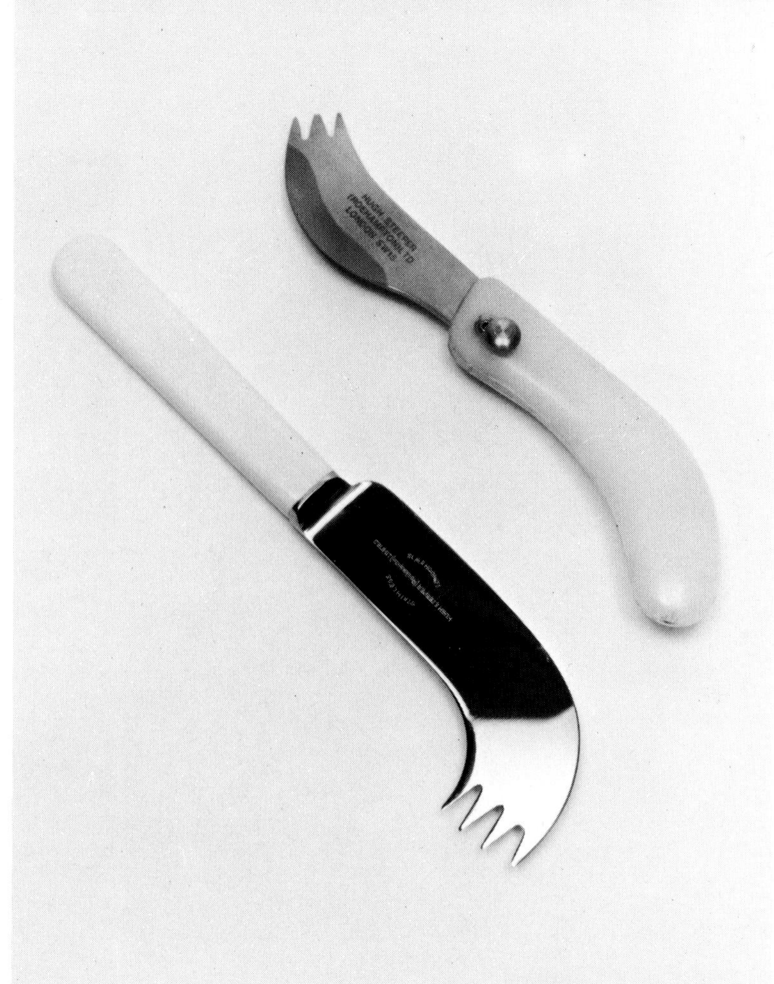

Product: Nelson Knife
(This is a combination of knife and fork for use when function of one hand is impaired.)
Client: Roehampton
London, England
Courtesy: The Able Child
New York, New York
Materials: Stainless steel and plastic

Product:	Multigrip cutlery with handle system (This system is for those with grasping or arm movement difficulties.)
Client:	Roehampton London, England
Courtesy:	The Able Child New York, New York
Materials:	Stainless steel and plastic

Product:	Eat and Drink Cutlery (This is designed for persons with poor grip or impaired hand/arm mobility)
Design Firm:	Ergonomi Design Gruppen AB
Client:	RFSU Rehab, Stockholm, Sweden
Materials:	Flatware: black polycarbonate handles and stainless steel; wine glass with extra broad foot: polycarbonate; plate with raised-lip edge and rubber friction ring to prevent sliding: white heat-resistant melamine

Design Awards Programs

ASID International Product Design Competition
American Society of Interior Designers
1430 Broadway
New York, New York 10018
USA

The Australian Design Award
Industrial Design Council of Australia
37 Little Collins Street
Melbourne 3000
Australia

Braun Prize for Technical Competition
Braun AG
Informationsabteilung
Postfach 1120
6242 Kronberg
West Germany

Design Council Awards
The Design Council
28 Haymarket
London SW1Y 4SU
England

Design Review
Industrial Design Magazine
330 West 42nd Street
New York, New York 10036
USA

Dunlopillo Design Award
Dunlop Limited
Dunlop House
Ryder Street
London SW1Y 6PX
England

IBD Product Design Competition
Institute of Business Designers
1155 Merchandise Mart
Chicago, Illinois 60654
USA

ID Prize
The Danish Design Council
H.C. Andersen Blvd., No. 18
1553 Copenhagen V
Denmark

IDEA Industrial Design Excellence Award
Industrial Designers Society of America
6802 Poplar Place
Suite 303
McLean, Virginia 22101
USA

International Design Competition
Japan Design Foundation
Semba Center Bldg. No. 4
Higashi-ku, Osaka
541 Japan

Pradikat if Die gute Industrieform
Die gute Industrieform
Hannover e. V.
Messegelände
3000 Hannover 82
West Germany

Roscoe Annual Product Design Award
The Resources Council Inc.
979 Third Avenue
Room 902, North
New York, New York 10022
USA

Shell Design Awards
Design Institute
South African Bureau of Standards
Private Bag X 191
Pretoria 0001
South Africa

Signe D'Or Exposition
Design Centre asbi
Galerie Ravenstein 51
1000 Bruxelles
Belgium

SMAU Industrial Design Award
Corso Venezia 49
Milan, Italy

Stuttgart Design Selection
Design Center Stuttgart
Landesgewerbeamt
Baden-Wurttemberg
Kienastrasse 18
7000 Stuttgart 1
West Germany

Products

Aircraft
  AVTEK 400, 187, 198
  Boeing 767 interior, 199
  Westland 30 Series helicopter, 200
Airless I MK II line marking machine, 203
Airmail Travel Hair Dryer, 39
Alistro, 77
Allegroh, 34
Alpino Pamir III isothermic tunnel tent, 211, 215
Amerec 610 Precision Rowing Machine, 211, 212
Andover Chair, 116
Appliances, 10-11, 30-43
Armor Cloth, 228
Aton Fluorescent, 78
Audioscope™, 184
Aurora Borealis, 65, 66, 68
Austopole rebutting machine, 209
Australia II keel, 221
Automobile equipment
  co-driver vehicle control device, 243
  Dynamic classics auto vacuum, 36
  JBL auto speakers, 50

Bed, The Capsule hotel, 151
BEGA downlight, 64
Bicycle, portable, 219
Bistro tempered teapot and glasses, 19
Blackboard, "Lili" writing table and (for children), 137
Blends (textiles)
  County Check, straw, 231
  County Downs, straw, 231
  Jhane Barnes Collections, 235
  Mirage Series, 229
  Ribcord, moonstone, 231
  Prestwick Cloth, 234
  Silhouette, 231
  Tapestry Cloth, 234
  Wool Andes, 228
Botta Chair, 129
Bottle opener, 14
Braun Calculator, 43
Briefcases
  American Tourister Business Equipment, 17
  Challenge Case and Briefcase, 16
  Swedish attache case, 17

Business equipment, 13, 17, 60, 61, 152-171. *See also* Furnishings, contract & residential
Button Magic Button Sewer, 31

Cabinet, classical, 139
Cabriolet, 70
Cafe-Bar Services, 169
Capsule, The (hotel bed), 151
Cardiac strobe, 184
Carpet
  California, 227
  Diamond Dancer, 235
  Stripes, 230
Cart-Mobile, 141
Casserole, 18
Chairs. *See also* Handicaped, designs for; Office seating; Wheelchairs; *specific kind of*
  Argyle (1897), 124
  Cabaret, 146
  Carlos Riart Chair, 127
  D. S., 2-4 (1918), 124
  Everychair Series, 123
  Experiment Collection, 147
  First, 146
  Flip-Seat™, 112-113
  Grid Chair, The, 116
  Hill House 1 (1902), 124
  Imperiale, 146
  Inna, 126
  Kita Collection, 120-121
  Klassik & Klasse, 115
  lounge (1929), 126
  MartinStoll Collection/G, 118
  Michael Graves Seating Collection, 145
  public, prototype, 111
  Red and Blue (1918), 125
  Richard Meier Collection, 149
  Royal, 147
  stacking, 122
  Steamer Collection, 128-129
  Transat Armchair (1927), 125
  Trapezoid, 148
  Willow 1 (1904), 124
  Willow 2 (1904), 125
  wire structure combinations, 110-111
  Zig-Zag (1934), 125
Champagne flutes, frosted, 27
Chevrolet Corvette (1984), 187, 188, 190
Children, design for
  "Lili" writing table and blackboard for children, 137
  Natasa Child's Cot, 136
  ORLAU Swivel Walker, 242
  walker, 241
  wheelchair, 241
China and pottery
  Ambassador, 22
  Culinaria, 23
  Pfaltzgraff Collection (Marimekko), 23, 26
CHRONOGYR room thermostat for small heating systems, 32
Clocks
  Radius Two Collection, 43
  Time of the Earth, 42
  Vacform wall clock, 42
CMX Systems videotape editor console, 165
Coatron Coagulometer, 185
Computed tomography system, CT9800, 172, 173, 179
Computers
  CAD system prototype, 160
  CIE-7800, 158
  Compact Computer, 40, 158
  Compass, 156
  Digital 350 Professional Series, 49, 61
  home prototype, 60
  IBM PCjr, 60
  IBM System 23, 153, 155
  ITT 3840 work station and disc module, 157
  M 10 portable, 157
  "The Machine of the Year" (prototype), 60, 61
  MAD-1, 153, 154
  modular school, system, 160
  PCS-Cadmus 2200 terminal, 154
  Phaze P3278 computer terminal, 155
  Pronto Series 16, 154
  StacPac Modules, 156
  VME/10, 158
  Wang Professional, 159
  WorkSlate, 159
  WY-1000 terminal, 157
Coordinated Resources Program, 222, 231, 232, 233
Copiers, paper
  Beta 4502, 165
  Oce 1725, 164
  Oce 1900, 164
  Oce 3760 Electro Static A1 Micro Enlarger/Printer, 165
Corning Model 102 printer, 173, 175
Cot, Natasa Child's, 136
Cotton textiles
  Fret, 225
  Kinnasand Collection, 229
  Maisema, 226
  Mira Columba, 226
  Monochrome Range of Curtaining, 235
  Night Hawk™, 226
  Orion™, 224
  Scroll, 225
  Shibumi, 224
  Tracery, 225
  Unique Cloth, 234
  Valkea Yo, 227
Courthouse Chair, 115
Crown Series TS Turret Sideloader, 187, 207
Cruet set, 29
Cupboard, corner, 142
Cutlery
  Eat and Drink Cutlery, 245
  Multigrip, with handle system, 245
  Nelson Knife, 244
Cyclos, 69
Desk(s). *See also* Office systems
  Gwathmey Siegel Desk, 99
  Master, 108
  Nokia personal computer and workstation, 171
  Pinstripe Office Furniture, 101
  SK-7 Desk, 101
  Trading Desk, 102
Desk accessories, 12, 13
Digital pH meter, 173, 174
Door, 142
Driv-Lok Grooved Pin, 175
Dunlop Max 200G mid-head graphite tennis racket, 213
Dynamics Classics, exercise tool, 213

Eclipse Saddlepack, 214-215
Electrophoresis densitometer, 176
Elevator cab, 142
Eltron shavers, 38
Entertainment, home, 48-61
Essleele "Australis," 18
EVAC + Chair, 238

Fabric, *see* Textiles
Fire extinguisher, home, 33
Fireplace tools, 35
Flame Photometer 943, 181
Flatware
  Century, 15
  Design 10 Plastic, 15
  Maya, 14
Flip-top Durabeam, 32
Fonar Beta 3000 Permanent Magnet, 178
Force Fin, 210, 214
Ford Sierra, 192
Forklift-truck, explosion-proof (prototype), 209
Fruit tray, 23
Furnishings, contract & residential, 88-148

G Series frontlift truck, 204-205
Garden seat, 148
Gibigiana, 73
Gina Chair, 115
Glassware, 29
  Bistro glasses, 19
  frosted Champagne Flutes, 27
  Reticelli, 21
Grand, 80
GSM Taxi, 195

Halogen Torchiere, 81
Halo Track Lighting, 84
Handicapped, designs for, 236-245
Hand Prehension Stimulator FESE H3, 183
Heater, fan, 33
Helicopter, Westland 30 series, 200
Hematrak™ Geometric Corporation, 176
Home electronics and entertainment, 48-61
Housewares, 10-29
HUP-180 portable hydraulic pump, 202
Hypo-Count II blood glucose monitor, 176

IMP II improved Mobility Package, 239
Industrial equipment, 186, 201-209
Isicom Super, 171
Iskrascope LCD, 159

Juice jugs and cups, 28

K-Line cigarette lighter, ash trays, and coasters, 25
Kongskilde, rotocrat, 201
Kroy laminator, 161

Lamps, *see* Lighting
Lance ski suit, 211, 220
Laser, handheld, 168
Latis chair, 144
Ledu, IPL 600 assymetrical task light, 86
LetraGraphix typesetter, 161
Library Laser, 167
Lighting, 62-87
Lighting installation, 87
Lotek Floor/Table Lamp, 83

Magazine rack, 34
Mary (single tube) and Mary Sue (double tube), 66-67
Medical equipment, 134, 172-185
Micronunciator alarm annunciator, 166
Micropad, 170
Mirror, Greek Revival, 143
Mixer tap, 35
Modem and chronograph, 170
Motorcycle, foldable urban, 218

NCR 2126 retail checkout system, 160
Neon sconce, 71
Neon torch, 82
Nike cross country ski racing boots, 210, 216-217
NMR Teslacon™ System, 173, 178
Norelco MicroSeries Microcassette Recorder/ Transcriber MCR-7200, 162

Odontosurge, 182
Office seating. *See also* Chairs
- Balans Vital 6000, 103
- Balans Vital 6035, 103
- Dorsal™ Seating Range, 89, 105
- executive, 108
- Helena Chair, 89, 106-107
- Kevi™ Chair, 105
- Kinetics Business Seating, 109
- Sirkus Chair, 107
- Stephens Office Seating, 104

Office systems
- ACM component system, 89, 91
- Action Office, 92
- Alpha Office System, 100
- COM System, 93, 94
- Data Entry Work Station, 99
- Dolmen office furnishings, 98
- Dunbar S/4 Series, 90
- Ergodata Office System, 89, 96-97
- executive desk system, 95
- general desk system, 95
- reception desk system, 95

Omeg/Balco digital insulation and continuity tester, 166
Orbis, 70
Oseris Spotlight Range, 85

Paging machines
- Monarch 120 Call-Connect, 163
- Sensar FM, 162
- Teletracer 2800, 163

Pausenia, 77
Pen(s)
- Safari pens, 13
- Lamy White Pens, 13

Penelope Chair, 119
Pherotron modular densitometer system, 182
Pillow speaker, 181
Pitcher, executive thermo, 28
Point-of-Scale terminal, 162
Polifemo, 79
Polyart abstract pattern jigsaw, 244
PO_2rtable Oxygen Monitor, 177
Post Card Travel Iron, 41
Pottery, *see* China and pottery
Projectorlite Night Vision Aid, 86
Protocol convertor, 171

"Quick Drop" hopper, articulated 3-axle, 208

Radios
- Sony Sports Walkman, 48, 56
- Watchman, FD-20A, 48, 56
- Watchman, FD-30A, 48, 56

Radius Two Collection, 12, 43
Razor, 39
Recreational and sports equipment, 210-221
Road tanker, 206
RO-BOOM, Roulund floating oil boom, 186, 202-203
Roll-In Shower, 242
Rug, *see* Carpet
Ryba, 143

Safety cushion prototype, 217
Samite electrical protection device, 167
Schultz Chair, 120
Serengeti Sunglasses, 40
Sewing machines
- Domestic sewing machine, 30
- Logica electronic sewing machine, 30
- Match-A-Patch Hole and Tear Mender, 31
- Singer Easy menders, 31

Shower stall, seat for, 244
Silk
- Isadora™, 227
- Metro, 233

Sintesi Track Lamp, 83
Slim chair, 128
Sofas
- Cabaret, 146
- LC 2 (1928), 124
- Michael Graves Seating Collection, 145
- Sinbad, 147

Solar screen, 150
Sports equipment, *see* Recreational and sports equipment; *specific kind of*
Stadium seat prototype, 221
Statsep plasma separator, 185
Steinberger Bass, 57
Stereo equipment
- ADS Alelier Audio Components, 48-49, 55
- Apt 1 Stereo Power Amplifier, 53
- Beocenter 7700, 54
- Concept 51 K, 52
- JBL, auto speakers, 50
- Kyocera DA-01 Compact Disc Player, 53
- MCS Stereo Receiver 3285, 50
- MCS Turntable 6730, 51
- MCS 1210 Stereo System, 52
- Technics SL-QL 15 turntable, 48, 51
- XRM-10 Music Shuttle, 55

Stitch-Me-Quick Hem and Seam Mender, 31
Stoneware, *see* China and pottery

Tables. *See also* Office systems
- adjustable-height, 135
- Balans Activ-6000, 103
- Bistro Table, 133
- Broken Length, 141
- Cinquecento, 145
- coffee, 130
- conference, 130
- D. S., 2-4 (1918), 124
- 45 Table, 132
- hospital bedside, 134
- "Lili" writing, and blackboard for children, 137
- L System, 141
- Lucia Mercer Collection, 135
- MartinStoll Collection/G, 118
- Modern Post-Neo, 140
- Neapolitan, 139
- Pina Gamba, 131
- Post-Box, 136-137
- Strata, 138

Tea and Coffee Service, 20, 27
Teapots, 19
Telemate, 180
Telephones
  Disa Telephone 1200 Type Range, 168
  ETA 80 telephone set, 43
Teleport 9 professional walkie-talkie, 167
Televisions
  FROGLINE Modules, 59
  Microtek 36cm remote control portable, 58
  Seiko Pocket Color, 58
Tellermate currency counter, 170
Temple Chair, 138
Tete-a-Tete, 140
Textiles, 222-235. *See also* Blends (textiles); Carpet; Cotton textiles; Silk; Wool textiles
Thermometer, electronic with LCD, 177
Tiny Tailor Mending machine, 31
Tokio, 71
Tools, 10-11
  electric drill, 44, 46
  pocket socket, 46
  Toro Compact 50 horse/reel system, 47
  Wagner Power Roller, 44
  woodworking safety kit, 45
  Torchiere, 80, 81
Toyota Van (1984), 187, 193
Tractor, towing, for forestry, 208
Trains
  Gas-Trac combustible gas detector, 196
  LRC interior, 194
  Transrapid 06 magnetic levitation, 186, 187, 196-197
Transfer Tub Seat, 243
Transportation, 186-209
Trucks, *see* Industrial equipment
Tupperware Modular Mates™ Container System, 24

Uni chair, 117

Vacuum cleaners
  Cyclon vacuum cleaner, 11, 37
  Dynamic Classics auto vacuum, 36
  Eureka Mighty Mike, 36-37
Van, Toyota (1984), 187, 193
Video equipment
  CMX Systems videotape editor console, 152, 165
  Record 2 video detail booster, 169
  Y-688[32] total error corrector video, 168

Walkers
  child's, 241
  ORLAU Swivel Walker, 237, 242
Wheelchairs
  child's, 241
  electric curb-climbing, 239
  foldable standard, 238
  restraint system for, 237, 240
  Titann Wheelchair, 237, 241
Wholly Cow (leather), 230
Wool textiles
  Classic Woolens, 228
  Clockwise, 232
  Chroma, 232
  Diamond Jubilee, silverweed, 231
  Hue, 232
  Limerick, meadow, 231
  Merrion Square, silverweed, 231
  Northern Lights, 232
  Nova, 232
  Stellar Collection, 232
  Sundown II, agate, 231
  Waffle Weave, 234

Designers

Aalto, Alvar, 68, 69, 78, 79
Akiyama, Osamu, 42
Aldersley, Michael, 206
Allen, Davis, 116
Allen, Trip, 210, 216
Altchek, Henry, 14
Ambasz, Emilio, 85, 89, 105, 140
Andrews, H.F., 167
Ann Maes Industrial Design, 34, 35
Arvinte, Mircea, 221
ASC Inc., 239
Ash, Bill, 45
Atelier International, 73

Bachmann, Urs, 97
Bahnsen, Uwe, 192
Baines, P., 167
Ballone, Michael, 46
Balmford, D.E.H., 200
Bang & Olufsen A/S, 54
Barbu, Nicolae, 221
Barnes, Jhane, 235
Barrier Free Environments, Inc., 243
Bartlett, Stephen, 22
Bash, Bill, 237, 241
Baumgarti, N., 167
BEGA, 64
Bellinger, Gretchen, 227
Bennett, Ward, 116, 141, 234
Bertok, Biba, 136
Best, Melvin, 219
Bets, B.A., 168
Byere, Eric, 184
BIB Design Consultants, 22, 32, 46, 170
Biondi, C., 93, 94
Blanchard, Rick, 50, 51, 52
Boehm, Michael, 21
Boeing Commercial Airplane Company, 199
Boris Kroll Design Studio, 232
Bostrum, James, 236, 242, 244
Botta, Mario, 129
Brammer, P., 200
Braun AG, 11, 33, 38, 43
Brooks, Steven, 90
Buchholz, John F., 175

Campbell, John Willy, 75
Campbell, Mark, 134, 135
Campetti, G., 167
Carlyle, Anne, 134, 135
Carne, Francis, 217
Carr, M.A., 58
Casciani, Stefano, 146
Cassia, Antonio Macchi, 157
Castelli, Clino, 93, 223
Castiglioni, Achille, 73, 146
Cavaliere, Alik, 145
Celentano, Linda, 23

Celine, D., 166
Certucha, Antonio Ortiz, 218
Chadwick, K., 58
Charles Pollock Inc., 119
Chelsea, Dan, 71, 82
Chiasson, Paul, 139
Clives, S.W., 167
Cohen, N., 243
Coloni, Tatjana, 137
Consumer Design Center, 158
Control Logic (Pty) Ltd., 166
Coons and Beirise Design Associates, 180
Corning Medical/Industrial Design Dept., 174
Coronelli, M., 70
Corporate Design Center, 45
Corporate Industrial Design, NCR Corporation, 160
Cunningham, C., 58

Dair, Tom, 40, 41
Dale E. Fahnstrom Design, 196
Danko, Peter, 123
Dave Kelley Design, 154
David A. Mintz, Inc., 67
Davin Stowell Associates, 23, 39, 40, 41, 161
Davison, L., 235
Dawson, Robert, 31, 44
Day & Ernst, 235
Dayton, Douglas C., 159
DCA Design Consultants, 163
De Licio, L., 70
dePolo, Lydia, 90
dePolo/Dunbar, 90
Derbyshire, R., 168
Design/Joe Sonderman, Inc., 201
Designspring, Inc., 86
Dettinger, Ernst, 115
deVitry, John, 142
Diffrient, Neils, 89, 107
Digital Equipment/Industrial Design Group, 61
Dittert, Karl, 91
Doe, R.A., 200
Douglass Ball, Inc., 237, 239, 240, 241
Driv-Lok, Inc/staff design, 175
Drummond, A.S., 167
Duff-Norton Company, 201
Dunbar, Jack, 90
Dwinger, Horst, 240
Dyson, James, 37

Eckhoff, Tias, 14
Eclipse Inc., 215
Edelstein, Matt, 80
Egen, David, 238
Egen Polymatic Corporation, 238
Ellefson, Clark, 138
Engineered Plastic Products, 161
Ergonomia Design, 171
Ergonomi Design Gruppen AB, 245
Evans, Bob, 210, 214
Evanson, James, 144, 148

Fabian, Wolfgang, 13
Faucheux, Brian, 138
Favaretto, Paolo, 95, 109
Fellmann Design AG, 32
Ferrieri, Anna Castelli, 133
Field, Brian, 206
Fischer-Design, 155, 157, 158, 162
Forcolini, Carlo, 79
Ford of Europe Inc., 192
Ford Motor Company, 239
Formosa, Daniel, 40, 41
Fortel Inc., 168, 169
Frank O. Gehry and Associates, 143
Frascaroli, Francesco, 93, 94
Frogdesign, 16, 34, 52, 59, 100, 167
Frost, Geoffrey, 131
Frost Design, 131
Fullwood, Maurice, 206
Furniture Club, 72, 130

Gamborini, Gino, 98
Gans, M., 166
Geer, Ken, 210, 216
Gehry, Frank, 143
General Electric Medical Systems Operations, 179
General Motors Design, 188
Gismondi, Ernesto, 77, 78, 83
Giugiaro, Giorgetto, 31
Glaser, Milton, 140
Glassel, Brian, 203
Golkin, Carrie Goldwater, 224, 226
Golkin, Gary, 224, 226
Goodwin, David, 112
Goodwin-Wheeler Associates, 112
Goof, Lennart, 182
Grange, Kenneth, 39
Graves, Michael, 145, 225
Gray, Eileen, 88, 125
Green, Patricia, 224
GSM Design Inc., 194, 195
Gusrud, Svein, 103
Gwathmey, Charles, 99

Haas, Chuck, 45
Hacker, Chris, 52
Haigh, Paul, 110
Haigh Architecture & Design, 110
Hakala, Pennti, 126
Halsted & Muramatsu, 50
Hannah, Bruce, 99
Hansen, K.G., 15
Hardy, Tom, 155
Hartiani, K., 168
Hayward, Jim, 95
Heddebaut, Andre, 215
Heininger, Oskar, 184
Herbst, Rene, 126
Hesse, Wolfgang, 209
Hill-Rom Company, 180
Hohulin, Samuel E., 37
Holdstock, Paul, 206
Holgersen, H.O., 202
Hollein, Hans, 20
Huldt, Johan, 128
Humphrey, G.H., 200
Hurley, Nancy, 243
Hypoguard Ltd./staff design, 176

IBM Entry Systems Division, 60
Ideograma S.C., 218
ID Two, 153, 154, 156
Ikegami, Toshiroh, 111
Imanaka, David, 212
Industrieel Ontwerpburo Berkheij, 163
Innovations & Development, Inc., 46
Instrumentation Laboratory Inc., 176, 177
Ishimoto, Fujiwo, 226, 227
Iskra Kibernetika, 159

Jack Lenor Larsen design studios, 231
Jackson, W. Shaun, 215
Jahn, Helmut, 143
Jensen, Finn Ulrich, 201
John Clark Shopsmith Inc., 45
Johnson, Gary C., 33
Junker, K.W., 166

Keith Muller Limited, 122, 134, 135
Kemker, Uwe, 160
Kennedy, Shane, 130
King, Miranda, Armaldi, 65
King, Perry A., 157
King Casey Inc., 31, 44, 47
Kisho Kurokawa Architect & Associates, 151
Kita, Toshiyuki, 121
Klojcnik, Ljuban, 159
Knighton, Charles, 192
Kruse, Erich, 209
Kuba, Lawrence M., 159
Kuhn, C., 235
Kukkapuro, Yrjo, 107, 147
Kurokawa, Kisho, 151
Kurokawa, Masayuki, 25
Kusmer, Ray, 84
Kyocera International Inc. (engineering division), 53

Lamb, Thomas, 129
Laude, Michael, 31
Lawing, Ed, 17, 61
Leby, Edward, 46
Le Corbusier, 88, 124
Lee Manners and Associates, 52
Lee Payne Associates, Inc., 139, 170, 171
Lefler, Al, 201
Levy, Edward, 46
Lewis, David, 182
Lewis, Jim, 138
Lewis & Clark, 138
Lexcen, Ben, 221
Lindau & Lindekrantz, 137
Lonczak, John, 39, 41
Lowes, B., 205
Lucchi, Michele de, 69, 80, 146
Lucker, Louis, 164, 165

McDonald, Roger, 230
Mace, Ronald L., 243
Mackintosh, Charles R., 124, 125
Mackintosh, Rennie, 88
Maes, Ann, 34, 35
Magistretti, Vico, 147
Main, B., 200
Makulik, Bernd, 115
Malan, J., 86
Malassigne, Pascal, 236, 242, 244
Manners, Lee, 52
Mathis, Scott, 174

Matrix Product Design, Inc., 157, 159
Matsushita Electrical and Industrial Company of Japan, 51, 177
Meier, Richard, 149
Mellor, R.W., 192
Melvin Best Industrial Design, 219
Mendini, Alessandro, 20
Mengshoel, Hans Chr., 103
Mercer, Lucia, 135
Michael W. Young Associates, 162
Miller, David, 44
Milton Glaser Associates, 140
Minolta Engineering Department, 165
Montague, Ed, 201
Mooney, Al, 198
Moore, Charles, 142
Moore, Ruble, Yudell, 142
Morgan, Stephen, 42
Morris, David, 45
Morrison S. Cousins and Associates, 36, 213
Motorola Inc./Paging Division, 162
Muller, Keith, 122, 134, 135
Muramatsu, David, 50
Murphy/Jahn, 143
Murray, Matthew, 184

Neumeister, Alexander, 154, 186, 197
Neumeister Design, 154, 182, 185, 197
Nishioka, Toru, 13
Nissan Motor Co. Ltd., 191
Nissen, Richard, 18
Nurmesniemi, Antti, 148
Nusse, K., 167
Nutall, Mike, 153, 154, 156, 159

Oakley, Nicholas, 170
Oce Design Team, 164, 165
Okuno, Tadahide, 150
Opsvik, Peter, 103

P.D.C.C. Studio, 139
Panton, Verner, 226
Parker, Kenneth R., 37
Pasquier, Nathalie du, 147, 227
Payne, Lee, 139
Pendleton, Thomas, 31, 33
Penney, JC, 50, 51, 52
Penney & Bernstein, 80, 87, 102, 168, 169
Peter Danko & Associates, 123
Peter Ralph Design Associates, 205
Pezdirc, Vladimir, 35
Pharr, Bruce, 168, 169
Piretti, Giancarlo, 85, 89, 105, 140
Plummer, Darril, 201
Pollock, Charles, 119
Porter, Paul W., 159
Portoghesi, Paolo, 20
Posts and Telecommuncations, Department of, 168
Potts, Robert, 174
Precor, 212
Premsela, Benno, 83
Prins, P., 208
Prototypes, 37
Pulos, Arthur, 184
Pulos Design Associates, Inc., 184

Raath, J.A., 168
Ralph, Peter, 205
Rams, Dieter, 33, 38, 43, 49, 55
Rasmussen, L.B., 201
Ratia, Ristomatti, 23, 26
Ratzlaff, Jorg, 238
Ray Tinley Halo Lighting Division, 84
Rezek, Ron, 70, 74, 81, 103
Riart, Carlos, 127
Ribeiro, M., 167
Richard Nissen, A/S, 18
Richardson/Smith, 17, 49, 61, 187, 207
Rieser, Frank, 238
Rietveld, Gerrit T., 125
Rivers, F.G., 200
Rodriguez, Gustavo, 60
Roe, Leonard, 206
Roeanick, Edward, 212, 220
Roger Williams Associates, 168
Ron Loosen Associates, 154
Ron Rezek Lighting and Furniture, 70, 74, 81, 103
Roos, Scott, 84
Rose, G.K., 242
Rossi, Aldo, 27
Rozier, Charles, 53
Ruddy, Don, 72, 130
Rupp, Tom, 50

Sacherman, Jim, 157
St. Zamora, Dan., 221
Sams, Bernard, 176
Sangyo Design, Zip Co., Ltd., 160
Sapper, Richard, 19
Savnik, Davorin, 42, 171, 183
Scarpa, Afra, 109
Scarpa, Tobia, 109
Schambourg, Charles, 230
Schnell, John, 45
Schultz, Richard, 120
Schultze, Fred, 201
Siegel, Hazel, 228
Siegel, Richard, 99
Sigheaki, Asahara, 71
SITE, 142
Skidmore, Owings, and Merrill, 116
Sklaroff, William, 12, 43, 101, 167, 168, 176, 181
Sky, Alison, 142
Small, Robert, 206
Smith, Burns D., 212
Smith, David B., 212
Smith, Morley, 194, 195
Sonderman, Joe, 201
Sonneman, Robert, 81
Sony Tokyo Design Engineers, 55, 56
Sottsass, Ettore, 29, 77
Steinberger, Ned, 57
Steinhilber & Deutsch Associates, 152, 165
Stephens, Bill, 104
Stern, Rudi, 71, 82
Stone, Michelle, 142
Stowell, Davin, 23, 40, 41
Studio 80, 13
Studio Kvadraf, 35
Studio Nurmesniemi, 148
Stultz Randy, 101, 115
Sweeney, Paul, 184
Swinfield, R.E., 200

Tam, Roy S.W., 30
Taylor, William, 198
Technicare Corporation, 178
Telephone Manufacturers of South Africa, 168
Teubner, H.M., 166
Thaler, Martin, 52
Thole, L., 166
Thomas Pendleton King Casey Inc., 44
Thomsen, Tamara, 39, 41
Tigerman, Stanley, 140
Tigerman, Fugman, McCurry, 140
Toffoloni, Werther, 116
Toyota Motor Corp., 193
Tribbey, Jan, 50, 51, 52
Tsiaparis, A., 167
Tucny, Peter, 208
Tupperware Design Groups, 24

Umeda, Masanori, 23
Urquidi, Luis, 50
Usab, Karen L., 159

Veaudry, H., 58
Venturi, Rauch, and Scott Brown, 143
Viemeister, Tucker, 39, 40, 41
Vignelli, Lella, 141
Vignelli, Massimo, 141
Vignelli Associates, 141
Villalobos, Armando Mercado, 218
V'Soske, Doug, 230

Walker/Group, Inc., 67, 101, 115, 231, 232, 233
Wallance, Don, 15
Walter Dorwin Teague Associates, 199
Ward Bennett Design, 116
Wardle, W., 205
Whall, Nigel, 206
Wieland, Rudolf M., 44
William Skarloff Design Associates, 43, 101, 167, 168, 176, 181
Wilson, Mark, 84
Wines, James, 142
Winter, Koen de, 29, 65
Wirkkala, Tapio, 15
Worrell, W. Robert, 161
Worrell Design Incorporated, 161

Yves Gonnet Inc., 233

Zaccai, Gianfranco, 176, 181, 185
Zolch, Volker, 160

Clients

A/S L. Goof, 182
A/S Roulunds Fabriker, 202
ADS, Analog & Digital Systems, Inc., 55
AEG Telefunken, 167
Ahmans i Ahus AB, 137
Alberta Children's Hospital, 134, 135
Alessi, 11, 19, 29
Alessi S.p.A., 20, 27
Alpino, 215
Ambiant Systems Ltd., 129
Amerec Corp., 212
American Tourister, 17
Apt Corporation, 53
Arteluce, 65
Artemide Inc., 69, 77, 78, 79, 83
Art People, 224, 226
Atelier International, 116
Austpole Pty. Ltd, 209
Austral Engineering Works (Pty), 208
Avarte Oy, 107, 147
Avtek Corporation, 198

Bang & Olufsen A/S, 54
Bellinger, Gretchen, 227
Beylerian Limited, 133
BG Watersports, 214
Black and Decker Ltd, 46
Boeing Commercial Airplane Company, 199
Bond, Alan, 221
Boris Kroll Fabrics, 232
Braun AG, 33, 38, 43
Braun Electronic GmbH, 55
Brickel Associates, 234
British Telecom, 163
Burns, Philp & Company Limited, 169

C. Josef Lamy GmbH, 13
C.O.M., 93, 94
Cardiac Imaging Inc., 184
Cassina, 147
Castelli Furniture Inc., 98, 119
Chevrolet Motor Division, General Motors Corporation, 188
C ITOH, 158
CMX Systems, Orrox Corporation, 165
Colorcore™, 139-143
Conran's, 128
Control Logic (Pty) Ltd., 166
Convergent Technologies, Inc., 159
Corning Designs, 23
Corning Glass Works, 40
Corning Medical, 175
Corning Science Products, 174
Crown Controls Corporation, 207